W9-BYF-410

Collins

Night Sky ALMANAC

A STARGAZER'S GUIDE TO

2022

Storm Dunlop & Wil Tirion

Published by Collins
An imprint of HarperCollins *Publishers*
Westerhill Road
Bishopbriggs
Glasgow G64 2QT
www.harpercollins.co.uk

HarperCollins *Publishers*
1st Floor, Watermarque Building, Ringsend Road, Dublin 4, Ireland

In association with
Royal Museums Greenwich, the group name for the National Maritime Museum, Royal Observatory Greenwich, the Queen's House and *Cutty Sark*
www.rmg.co.uk

A catalogue record for this book is available from the British Library

ISBN 978-0-00-846988-7

10 9 8 7 6 5 4 3 2 1

Printed and bound by CPI Group (UK) Ltd,
Croydon CR0 4YY

If you would like to comment on any aspect of this book, please contact us at the above address or online.
e-mail: collins.reference@harpercollins.co.uk

facebook.com/CollinsAstronomy
@CollinsAstro

MIX
Paper from
responsible sources
FSC™ C007454

This book is produced from independently certified FSC™ paper to ensure responsible forest management.

For more information visit: www.harpercollins.co.uk/green

Contents

Introduction

The aim of this book is to help people to find their way around the night sky and to understand what is visible every month, from anywhere in the world. The stars that may be seen depend on where you are on Earth, but even if you travel widely, this book will show you what you can see. The night sky also changes from month to month and these changes, together with some of the significant events that occur during the year are described and illustrated.

The charts that are used differ considerably from those found in most astronomy books, and have been specifically designed for use anywhere in the world. A full description of how to use and understand the monthly charts is given on pages 36–39.

Sunrise, sunset and twilight

The conditions for observing naturally vary over the course of the year and one's location on Earth. Sunrise and sunset vary considerably, depending in particular on one's latitude. Sunrise and sunset times are given each month for nine different locations around the world. These places are shown in a **bold** typeface on the world map on pages 40–41. Sunrise and sunset times are given for the first and last days in every month, for these specific locations. Another factor that influences what may be seen is twilight at dusk and dawn. Again, this varies considerably with one's latitude on Earth. The diagrams on pages 251–253 show how this varies for the nine locations, which have been chosen to show the range of variation, rather than just for the importance of the places that have been included. The different stages of twilight and how they affect observing are also explained there.

Moonlight

Yet another factor that affects the visibility of objects is the amount of moonlight in the sky. At Full Moon, it may be very difficult to see some of the fainter stars and objects, and even when the Moon is at a smaller phase it seriously interferes with visibility if it is near the stars or planets in which you are interested. A full lunar calendar is given for each month and

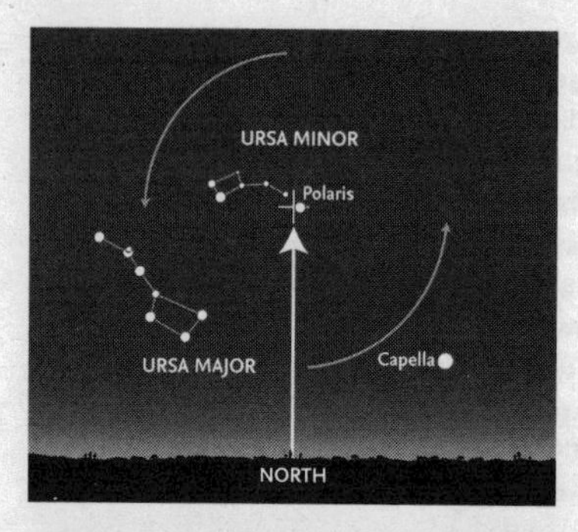

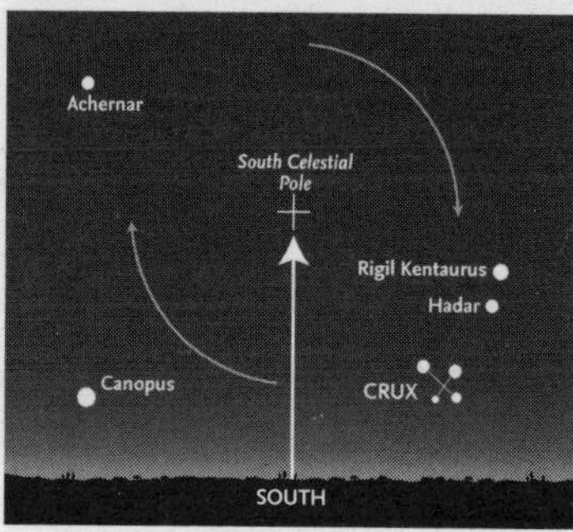

The altitude of the Celestial Pole equals the observer's latitude.

may be used to see when nights are likely to be darkest and best for observing.

The celestial sphere
All the objects in the sky (including the Sun, Moon, and stars) appear to lie at some indeterminate distance on a large sphere, centred on the Earth. This celestial sphere has various reference points and features that are related to those of the Earth. If the Earth's rotational axis is extended, for example, it points to the North and South Celestial Poles, which are thus in line with the North and South Poles on Earth. As shown in the diagrams, the altitude of the celestial pole is equal to the observer's latitude, whether in the north or south. Similarly, the celestial equator lies in the same plane as the Earth's equator, and divides the sky into northern and southern hemispheres.

It is useful to know some of the special terms for various parts of the sky. As seen by an observer, half of the celestial sphere is invisible, below the horizon. The point directly overhead is known as the ***zenith***, and this point is shown on the monthly charts for several different latitudes, where it is an important reference point. The (invisible) point below one's feet is the ***nadir***. The line running from the north point on the horizon, up through the zenith and then down to the south point is the ***meridian***. This is an important invisible line in the sky, because objects are highest in the sky, and thus easiest to see, when they cross the meridian in the south. Objects are said to transit, when they cross this line in the sky.

In this book, reference is sometimes made in the text and in the diagrams to the standard compass points around the horizon. The position of any object in the sky may be described by its ***altitude*** (measured in degrees above the horizon) and its ***azimuth*** (measured in degrees from north, 0°, through east, 90°, south, 180°, and west, 270°). Experienced amateurs and professional astronomers also use another system of specifying locations on the celestial sphere, but that need not concern us here, where the simpler method will suffice.

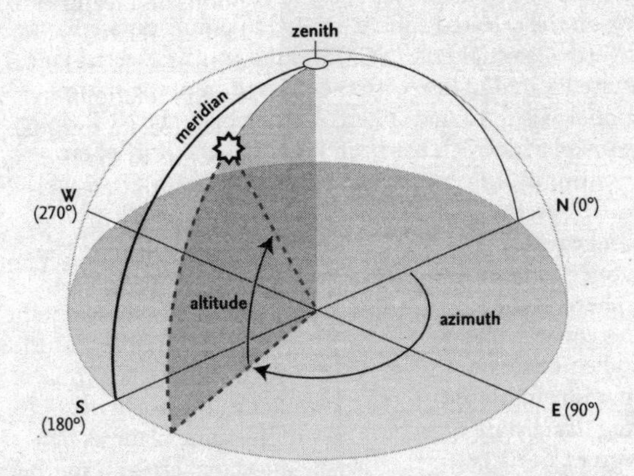

Measuring altitude and azimuth on the celestial sphere.

The celestial sphere appears to rotate about an invisible axis, running between the north and south celestial poles. The location (i.e., the altitude) of the celestial poles depends entirely on the observer's position on Earth or, more specifically, their latitude.

Right ascension and declination

On the previous pages we have mentioned how a simple method, involving altitude and azimuth, measured relative to the observer's horizon, may be used to specify the position of an object in the sky. Astronomers, however, use another, more precise method, which does not depend on the observer's position on Earth (and thus on their local horizon). This involves the two co-ordinates, ***right ascension*** (RA) and ***declination*** (dec). Right ascension is measured eastwards (to the left) from the ***First Point of Aries*** (page 78) in hours and minutes of time (and very occasionally, in seconds) or else (less frequently) in degrees. Because the Earth rotates once in 24 hours, one hour of right ascension equals 15 angular degrees. The sky appears to rotate by this amount in one hour.

All objects in the sky appear to be located on an imaginary sphere: the celestial sphere. There are, however, certain fixed points on the celestial sphere, related to points on the Earth. The North Celestial Pole (NCP) and the South Celestial Pole (SCP) are located in line with the projection of the Earth's rotational axis onto that sphere. In the north, the NCP is very close to Polaris, which has been known as the North Star since antiquity. In a similar way, the celestial equator is the projection onto the sphere of the Earth's equator. The second co-ordinate, declination, is simply the angular distance, in degrees, north or south of the celestial equator. The Sun has a declination of zero when it appears to cross the celestial equator at the equinoxes.

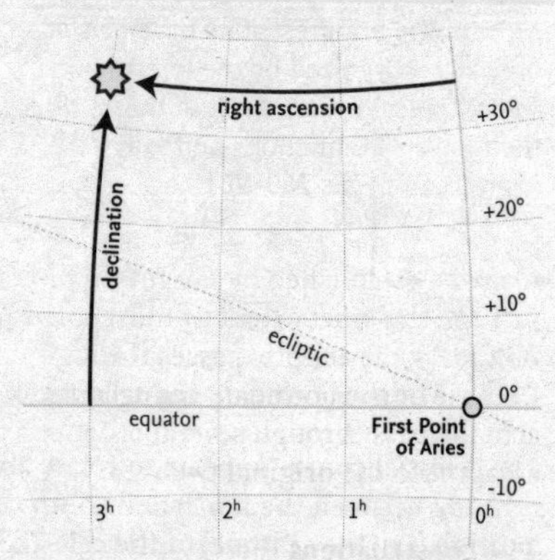

Measuring right ascension (RA) and declination (dec) on the celestial sphere.

The ecliptic and the zodiac

Another important line on the celestial sphere is the Sun's apparent path against the background stars – in reality the result of the Earth's orbit around the Sun. This is known as the ***ecliptic***. The point where the Sun, apparently moving along the ecliptic, crosses the celestial equator from south to north is known as the vernal (or northern spring) equinox, which occurs around March 21. At this time (and at the northern autumnal equinox, on September 22 or 23, when the Sun crosses the celestial equator from north to south) day and night are almost exactly equal in length. (There is a slight difference, but that need not concern us here.) The vernal equinox is currently located in the constellation of Pisces, and is important in astronomy because it defines the zero point for a system of celestial coordinates, which is, however, not used in this book.

The Moon and planets are to be found in a band of sky that extends 8° on either side of the ecliptic. This is because the orbits of the Moon and planets are inclined at various angles to the ecliptic (i.e., to the plane of the Earth's orbit). This band of sky is known as the zodiac, and when originally devised, consisted of twelve ***constellations***, all of which were considered to be exactly 30° wide. When the constellation boundaries were formally established by the International Astronomical Union in 1930, the exact extent of most constellations was altered, and nowadays the ecliptic passes through thirteen constellations. Because of the boundary changes, the Moon and planets may actually pass through several other constellations that are adjacent to the original twelve.

Most of the brightest stars have names officially recognized by the International Astronomical Union. A list of these, with their Bayer designations and magnitudes, is given on pages 260–261.

The constellations

In the western astronomical tradition, the celestial sphere has always been divided into various constellations, most dating back to antiquity and usually associated with certain myths or legendary people and animals. Nowadays, 88 constellations cover the whole sky, and their boundaries have been fixed by international agreement. Their names (in Latin) are largely derived from Greek or Roman originals. (A full list of the constellations is given on pages 256–258, with their names in English, their abbreviations, and genitive forms.) Some of the names of the most prominent stars are of Greek or Roman origin, but many are derived from Arabic names. Some bright stars have no individual names, and for many years, they were identified by terms such as 'the star in Hercules' right foot'. A more sensible scheme was introduced by the German astronomer Johannes Bayer in the early seventeenth century. Following his scheme – which is still used today – most of the brightest stars are identified by a Greek letter followed by the genitive form of the constellation's Latin name. An example is the Pole Star, also known as Polaris and α Ursae Minoris. The Greek alphabet is shown on page 258.

Asterisms

Apart from the constellations, certain groups of stars, which may form a small part of a larger constellation, are readily recognizable and have been given individual names. These groups are known as asterisms, and the most famous (and well-known) is the 'Plough' or 'Big Dipper', the common name for the seven brightest stars in the constellation of Ursa Major, the Great Bear. The names and identifications of some popular asterisms are given in the list on page 259.

The Moon

As it passes across the sky from west to east in its orbit around the Earth, the Moon moves by approximately its diameter (about half a degree) in an hour. Normally, in its orbit, the Moon passes above or below the direct line between Earth and Sun (at New Moon) or outside the area obscured by the Earth's shadow (at Full Moon). Occasionally, however, the three bodies are more-or-less perfectly aligned to give an eclipse: a solar eclipse at New Moon, or a lunar eclipse at Full Moon. Depending on

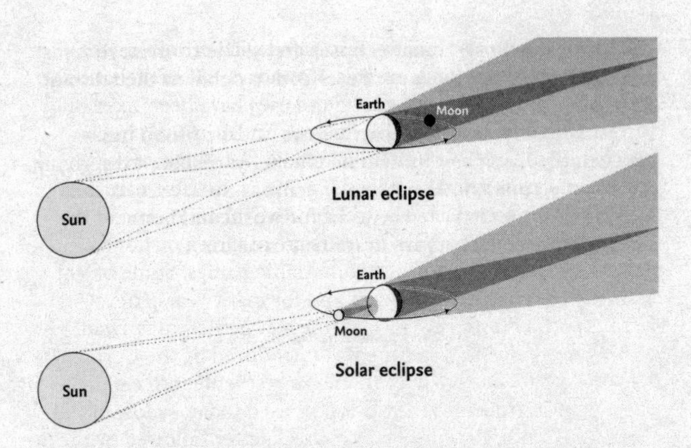

When the Moon passes through the Earth's shadow (top), a lunar eclipse occurs. When it passes in front of the Sun (below) a solar eclipse occurs.

the exact circumstances, a solar eclipse may be merely partial (when the Moon does not cover the whole of the Sun's disc); annular (when the Moon is too far from Earth in its orbit to appear large enough to hide the whole of the Sun); or total. Total and annular eclipses are visible from very restricted areas of the Earth, but partial eclipses are normally visible over a wider area. Two forms of solar eclipse occur this year, and are described in detail in the appropriate month.

Precautions must always be taken when viewing even partial phases of a solar eclipse to avoid damage to your eyes. Only ever use proper eclipse glasses, or a proper solar filter over the full objective of a telescope. The glass 'solar filters' sometimes provided with cheap telescopes should never be used. They are unsafe.

Somewhat similarly, at a lunar eclipse, the Moon may pass through the outer zone of the Earth's shadow, the penumbra (in a penumbral eclipse, which is not generally perceptible to the naked eye); pass so that just part of the Moon is within the darkest part of the Earth's shadow, the umbra (in a partial eclipse); or completely within the umbra (in a total eclipse).

Unlike solar eclipses, lunar eclipses are visible from large areas of the Earth. Again, these are described in detail in the relevant month.

Occasionally, as it moves across the sky, the Moon passes between the Earth and individual planets or distant stars, giving rise to an ***occultation.*** As with solar eclipses, such occultations are visible from restricted areas of the world, but certain significant occultations are described in detail.

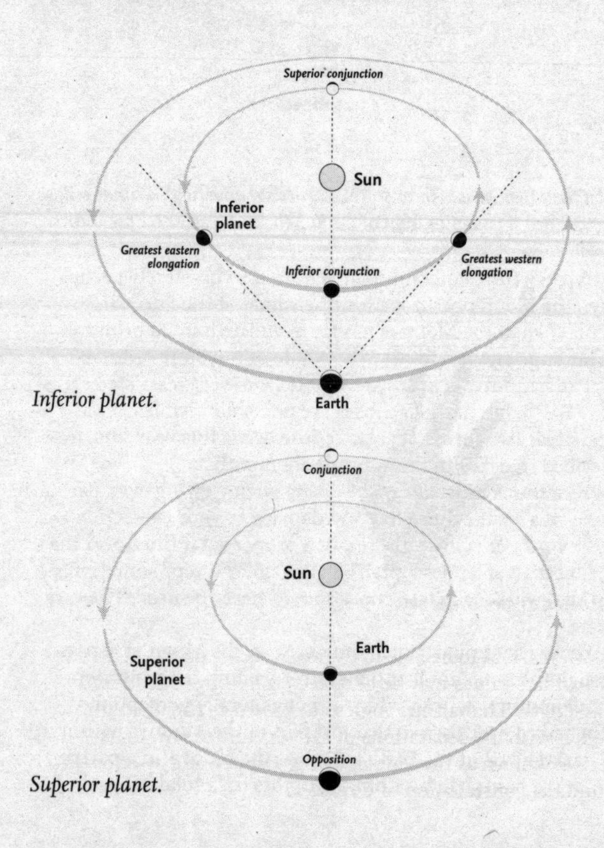

Inferior planet.

Superior planet.

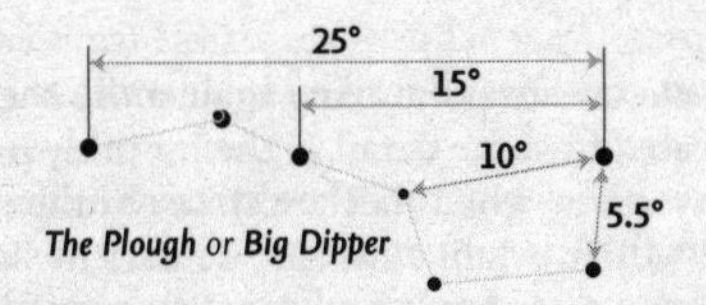

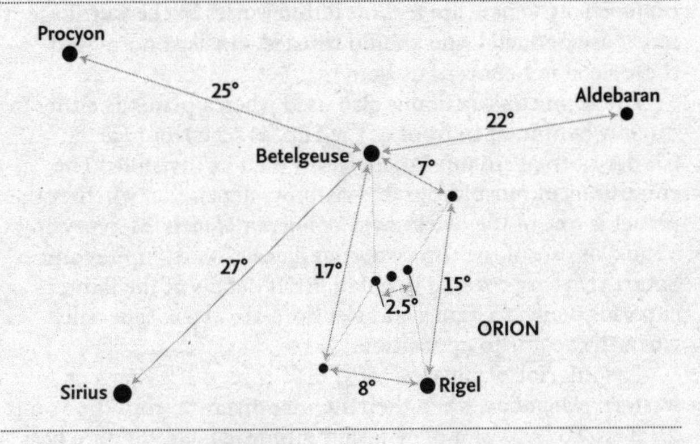

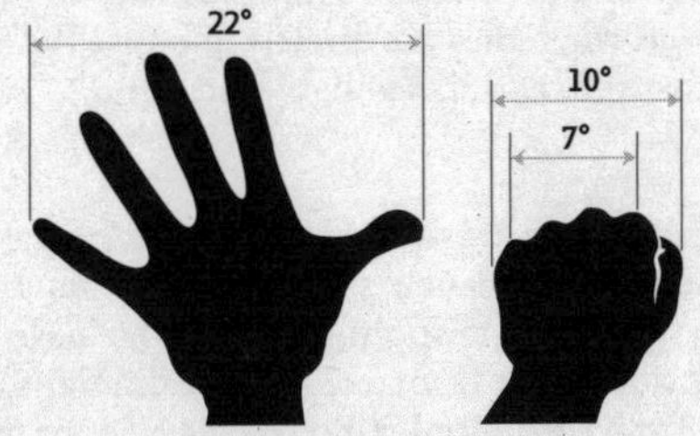

Measuring angles in the sky.

It is often useful to be able to estimate angles on the sky, and approximate values may be obtained by holding one hand at arm's length. The various angles are shown in the diagram, together with the separations of the various stars in the asterism, known as the Plough or Big Dipper, and also for stars around the constellation of Orion.

The planets

Because the planets are always moving against the background stars, they are treated in some detail in the monthly pages and information is given when they are close to other planets, the Moon, or any of five bright stars that lie near the ecliptic. Such events are known as ***appulses*** or, more frequently, as ***conjunctions***. (There are technical differences in the way these terms are defined – and should be used – in astronomy, but these need not concern us here.)

The term conjunction is also used when a planet is either directly behind or in front of the Sun, as seen from Earth. (Under normal circumstances it will then be invisible.) The conditions of most favourable visibility depend on whether the planet is one of the two known as ***inferior planets*** (Mercury and Venus) or one of the three ***superior planets*** (Mars, Jupiter and Saturn) that are covered in detail. Brief details of the fainter superior planets, Uranus and Neptune, are given, especially when they come to opposition.

The inferior planets are most readily seen at eastern or western ***elongation***, when their angular distance from the Sun is greatest. For superior planets and minor planets, they are best seen at ***opposition***, when they are directly opposite the Sun in the sky, and cross the meridian at local midnight.

Events

A number of interesting events are shown in diagrams for each month. They involve the planets and the Moon, sometimes showing them in relation to specific stars. Events have been chosen as they will appear from one of three different locations: from London; from the central region of the USA; or from Sydney in Australia. Naturally, these events are visible from other locations, but the appearance of the objects on the sky will differ slightly from the diagrams. A list of major astronomical events in 2022 is given on pages 20–21.

Meteors

At some time or other, nearly everyone has seen a ***meteor*** – a 'shooting star' – as it flashed across the sky. The particles that

Meteor shower (showing the April Lyrid radiant).

cause meteors – known technically as 'meteoroids' – range in size from that of a grain of sand (or even smaller) to the size of a pea. On any night of the year there are occasional meteors, known as ***sporadics***, that may travel in any direction. These occur at a rate that is normally between 3 and 8 in an hour. Far more important, however, are ***meteor showers***, which occur at fixed periods of the year, when the Earth encounters a trail of particles left behind by a comet or, very occasionally, by a minor planet (asteroid). Meteors always appear to diverge from a single point on the sky, known as the ***radiant***, and the radiants of major showers are shown on the charts.

Meteors that come from a circular area, 8° in diameter, around the radiant are classed as belonging to the particular shower. All others that do not come from that area are sporadics (or, occasionally, from another shower that is active at the same time). A list of the major meteor showers is given on the next page.

Although the positions of the various shower radiants are shown on the charts, looking directly at the radiant is not the most effective way of seeing meteors. They are most likely to be noticed if one is looking about 40–45° away from the radiant position. (This is approximately two hand-spans as shown in the diagram for measuring angles on page 13.)

Meteor Showers

Shower	*Dates of activity 2022*	*Date of maximum 2022*	*Possible hourly rate*
Quadrantids	Dec. 28 to Jan. 12	Jan. 3–4	110
α-Centaurids	Jan. 31 to Feb. 20	Feb. 8	6
γ-Normids	Feb. 25 to Mar. 28	Mar. 14–15	6
April Lyrids	Apr. 14–30	Apr. 22–23	18
π-Puppids	Apr. 15–28	Apr. 23–24	var.
η-Aquariids	Apr. 19 to May 28	May 6	50
α-Capricornids	Jul. 3 to Aug. 15	Jul. 30	5
Southern δ-Aquariids	Jul. 12 to Aug. 23	Jul. 30	25
Piscis Austrinids	Jul. 15 to Aug. 10	Jul. 28	5
Perseids	Jul. 17 to Aug. 24	Aug. 12–13	100
α-Aurigids	Aug. 28 to Sep. 5	Sep. 1	6
Southern Taurids	Sep. 10 to Nov. 20	Oct. 10–11	5
Orionids	Oct. 2 to Nov. 7	Oct. 21–22	25
Draconids	Oct. 6–10	Oct. 8–9	10
Northern Taurids	Oct. 20 to Dec. 10	Nov. 12–13	5
Leonids	Nov. 6–30	Nov. 17–18	10
Phoenicids	Nov. 28 to Dec. 9	Dec. 2	var.
Puppid Velids	Dec. 1–15	Dec. 7	10
Geminids	Dec. 4–20	Dec. 14–15	150
Ursids	Dec. 17–26	Dec. 22–23	10

Some Interesting Objects

Messier IC / NGC	*Name*	*Type*	*Constellation*
—	47 Tucanae	globular cluster	Tucana
—	Hyades	open cluster	Taurus
—	Double Cluster	open cluster	Perseus
—	Melotte 111	open cluster	Coma Berenices
M3	—	globular cluster	Canes Venatici
M4	—	globular cluster	Scorpius
M8	Lagoon Nebula	gaseous nebula	Sagittarius
M11	Wild Duck Cluster	open cluster	Scutum
M13	Hercules Cluster	globular cluster	Hercules
M15	—	globular cluster	Pegasus
M22	—	globular cluster	Sagittarius
M27	Dumbbell Nebula	planetary nebula	Vulpecula
M31	Andromeda Galaxy	galaxy	Andromeda
M35	—	open cluster	Gemini
M42	Orion Nebula	gaseous nebula	Orion
M44	Praesepe	open cluster	Cancer
M45	Pleiades	open cluster	Taurus
M57	Ring Nebula	planetary nebula	Lyra
M67	—	open cluster	Cancer
IC 2602	Southern Pleiades	open cluster	Carina
NGC 752	—	open cluster	Andromeda
NGC 3242	Ghost of Jupiter	planetary nebula	Hydra
NGC 3372	Eta Carinae Nebula	gaseous nebula	Carina
NGC 4755	Jewel Box	open cluster	Crux
NGC 5139	Omega Centauri	globular cluster	Centaurus

Other objects

Certain other objects may be seen with the naked eye under good conditions. Some were given names in antiquity – Praesepe is one example – but many are known by what are called 'Messier numbers', the numbers in a catalogue of nebulous objects compiled by Charles Messier in the late-eighteenth century. Some, such as the Andromeda Galaxy, M31, and the Orion Nebula, M42, may be seen faintly by the naked eye, but all those given in the list here will benefit from the use of binoculars.

Apart from galaxies, such as M31, which contain thousands of millions of stars, there are also two types of cluster: open clusters, such as M45, the Pleiades, which may consist of a few dozen to some hundreds of stars; and globular clusters, such as M13 in Hercules, which are spherical concentrations of many thousands of stars. One or two gaseous nebulae, consisting of gas illuminated by stars within them are also visible. The Orion Nebula, M42, is one, and is illuminated by the group of four stars, known as the Trapezium, which may be seen within it by using a good pair of binoculars. A list of interesting objects is given on the previous page.

In 1781, ***Charles Messier*** (26 June 1730 – 12 April 1817) published the final version of his catalogue of 110 nebulous objects and faint star clusters that might be confused with comets. The objects in this catalogue are still known as the Messier objects and are always quoted as 'M' numbers.

Dates and time

Astronomers, worldwide, use a standardized method of expressing the date and time. This prevents confusion in comparing observations made by different observers. The various elements are given in descending order: year, month (three-letter abbreviation to prevent confusion over using numbers), day, hour, minutes, seconds. (In extreme cases, fractions of minutes or seconds may be used.) The date and time are that on the Greenwich meridian (GMT), and ignore any changes for Summer Time / Daylight Saving Time (DST), and any adjustments for local time at the observer's location. This standard is known as Coordinated Universal Time (UTC). In this book and in many others this is generally given as Universal Time (UT). All times given in this book are in UT.

To avoid problems over the changes involved in moving to and from Summer Time / Daylight Saving Time (and the complications over the beginning and end dates) and also the adjustments for local time, experienced astronomers set a (cheap) watch or clock to Universal Time and keep it that way. Smartphone users may use a simple world clock app, provided they lock it to the time (GMT) on the Greenwich meridian.

Similarly, the date given for an event is the date as it applies at the Greenwich meridian, i.e., in UT. Occasionally, this may differ from the date as given by your local time. An event that occurs (say) late in the night in Europe may seem to occur on the previous day to an observer to the west (such as in the USA), when local time is taken into account. This is another complication that is avoided by using the Universal Time standard.

In 1801, the Italian astronomer Giuseppe Piazzi discovered what appeared to be a new planet orbiting between Mars and Jupiter, and named it Ceres. William Herschel proved it to be a very small object, calculating it to be only 320 km in diameter, and not a planet. He proposed the name asteroid, and soon other similar bodies were found. We now know that Ceres is 952 km in diameter. It is now considered to be a dwarf planet.

Major Events in 2022

Jan. 03–04	Quadrantid meteor shower maximum
Jan. 04	Earth at perihelion
Jan. 07	Mercury at greatest elongation
Jan. 13	Minor planet (7) Iris at opposition
Feb. 08	α-Centaurid meteor shower maximum
Feb. 16	Mercury at greatest elongation
Mar. 14–15	γ-Normid meteor shower maximum
Mar. 20	Venus at greatest elongation
Mar. 20	Northern spring / Southern autumnal equinox
Apr. 22–23	April Lyrid meteor shower maximum
Apr. 23–24	π-Puppid meteor shower maximum
Apr. 29	Mercury at greatest elongation
Apr. 30	Partial solar eclipse
May 06	η-Aquariid meteor shower maximum
May 16	Total lunar eclipse
Jun. 16	Mercury at greatest elongation
Jun. 21	Northern summer / Southern winter solstice
Jul. 28	Piscis Austrinid meteor shower maximum
Jul. 30	α-Capricornid meteor shower maximum
Jul. 30	Southern δ-Aquariid meteor shower maximum
Aug. 12–13	Perseid meteor shower maximum
Aug. 14	Saturn at opposition
Aug. 22	Minor planet (4) Vesta at opposition
Aug. 27	Mercury at greatest elongation

Major events in 2022 (continued)

Sep. 01	α-Aurigid meteor shower maximum
Sep. 05	Minor planet (3) Juno at opposition
Sep. 16	Neptune at opposition
Sep. 23	Northern autumnal / Southern spring equinox
Sep. 26	Jupiter at oposition
Oct. 08–09	Draconid meteor shower maximum
Oct. 08	Mercury at greatest elongation
Oct. 10–11	Southern Taurid meteor shower maximum
Oct. 21–22	Orionid meteor shower maximum
Oct. 25	Partial solar eclipse
Oct. 25	Occultation of Venus
Nov. 08	Total lunar eclipse
Nov. 09	Uranus at opposition
Nov. 12–13	Northern Taurid meteor shower maximum
Nov. 17–18	Leonid meteor shower maximum
Dec. 02	Phoenicid meteor shower maximum
Dec. 07	Puppid Velid meteor shower maximum
Dec. 08	Occultation of Mars
Dec. 08	Mars at opposition
Dec. 14–15	Geminid meteor shower maximum
Dec. 21	Mercury at greatest elongation
Dec. 21	Northern winter / Southern summer solstice

The Moon

The Moon at First Quarter.

The Moon
The monthly pages include diagrams showing the ***phase*** of the Moon (see page 28) for every day of the month, and also indicate the day in the ***lunation*** (or ***age*** of the Moon), which begins at New Moon. The diagrams showing the Moon's phase are repeated for southern-hemisphere observers who will see the Moon, south up. Although the main features of the surface – the light highlands and the dark maria (seas) – may be seen with the naked eye, far more features may be detected with the use of binoculars or any telescope. The many craters are best seen when they are close to the ***terminator*** (the boundary between the illuminated and the non-illuminated areas of the surface), when the Sun rises or sets over any particular region of the Moon and the crater walls or central peaks cast strong shadows. Most features become difficult to see at Full Moon, although this is the best time to see the bright ray systems surrounding certain craters. Accompanying the Moon map on the following pages is a list of prominent features, including the days in the lunation when features are normally close to the terminator and thus easiest to see. A few bright features such as Linné and Proclus, visible when well illuminated, are also listed. One feature, Rupes Recta (the Straight Wall) is readily visible only when it casts a shadow with light from the east, appearing as a light line when illuminated from the opposite direction.

The dates of visibility vary slightly through the effects of ***libration***. Because the Moon's orbit is inclined to the Earth's equator and also because it moves in an ellipse, the Moon appears to rock slightly from side to side (and nod up and down). Features near the ***limb*** (the edge of the Moon) may vary considerably in their location and visibility. (This is easily noticeable with Mare Crisium and the craters Tycho and Plato.) Another effect is that at crescent phases before and after New Moon, the normally non-illuminated portion of the Moon receives a certain amount of light, reflected from the Earth. This ***Earthshine*** may enable certain bright features (such as Aristarchus, Kepler and Copernicus) to be detected even though they are not illuminated by sunlight.

Moon features

The numbers below indicate the age of the Moon when features are usually best visible.

Abulfeda	6:20
Agrippa	7:21
Albategnius	7:21
Aliacensis	7:21
Alphonsus	8:22
Anaxagoras	9:23
Anaximenes	11:25
Archimedes	8:22
Aristarchus	11:25
Aristillus	7:21
Aristoteles	6:20
Arzachel	8:22
Atlas	4:18
Autolycus	7:21
Barrow	7:21
Billy	12:26
Birt	8:22
Blancanus	9:23
Bullialdus	9:23
Bürg	5:19
Campanus	10:24
Cassini	7:21
Catharina	6:20
Clavius	9:23
Cleomedes	3:17
Copernicus	9:23
Cyrillus	6:20
Delambre	6:20
Deslandres	8:22
Endymion	3:17
Eratosthenes	8:22
Eudoxus	6:20
Fra Mauro	9:23
Fracastorius	5:19
Franklin	4:18
Gassendi	11:25
Geminus	3:17
Goclenius	4:18
Grimaldi	13–14:27–28
Gutenberg	5:19
Hercules	5:19
Herodotus	11:25
Hipparchus	7:21
Hommel	5:19
Humboldt	3:15
Janssen	4:18
Julius Caesar	6:20
Kepler	10:24
Landsberg	10:24
Langrenus	3:17
Letronne	11:25
Linné	6
Longomontanus	9:23
Macrobius	4:18
Mädler	5:19
Maginus	8:22
Manilius	7:21
Mare Crisium	2–3:16–17
Maurolycus	6:20
Mercator	10:24
Metius	4:18
Meton	6:20
Mons Pico	8:22
Mons Piton	8:22
Mons Rümker	12:26
Montes Alpes	6–8:21
Montes Apenninus	8
Orontius	8:22
Pallas	8:22
Petavius	3:17
Philolaus	9:23
Piccolomini	5:19
Pitatus	8:22
Pitiscus	5:19
Plato	8:22
Plinius	6:20
Posidonius	5:19
Proclus	14:18
Ptolemaeus	8:22
Purbach	8:22
Pythagoras	12:26
Rabbi Levi	6:20
Reinhold	9:23
Rima Ariadaeus	6:20
Rupes Recta	8
Saussure	8:22
Scheiner	10:24
Schickard	12:26
Sinus Iridum	10:24
Snellius	3:17
Stöfler	7:21
Taruntius	4:18
Thebit	8:22
Theophilus	5:19
Timocharis	8:22
Triesnecker	6–7:21
Tycho	8:22
Vallis Alpes	7:21
Vallis Schröteri	11:25
Vlacq	5:19
Walther	7:21
Wargentin	12:27
Werner	7:21
Wilhelm	9:23
Zagut	6:20

Map of the Moon

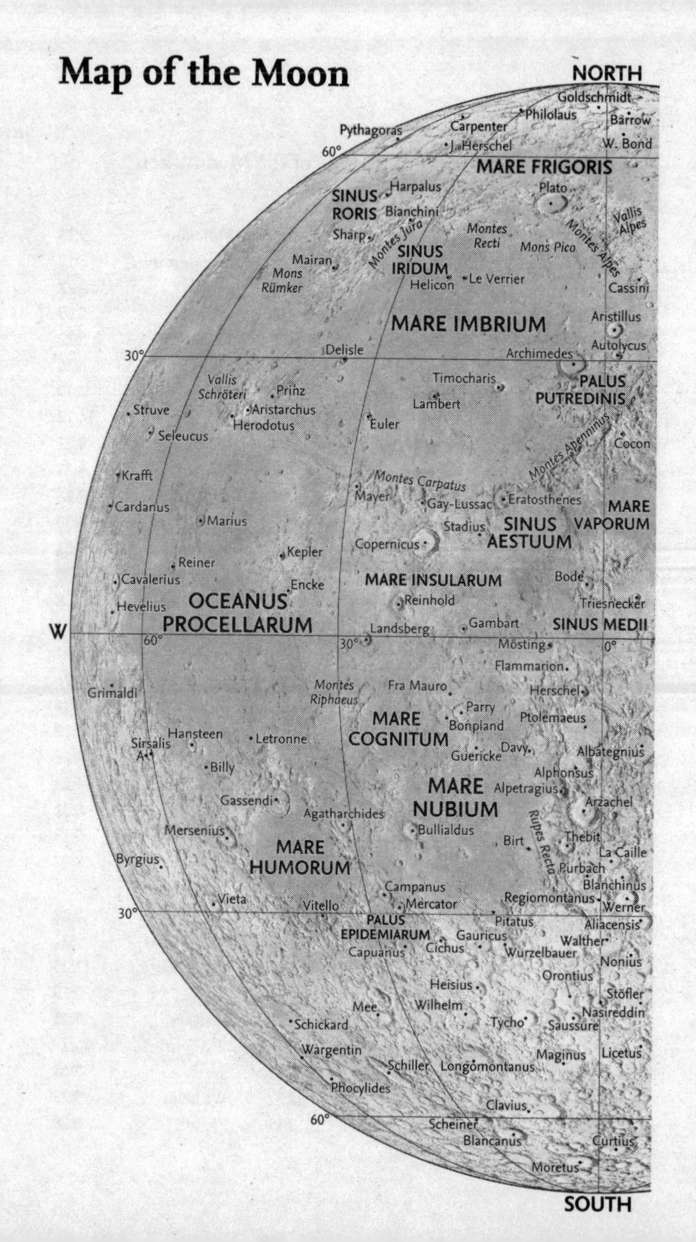

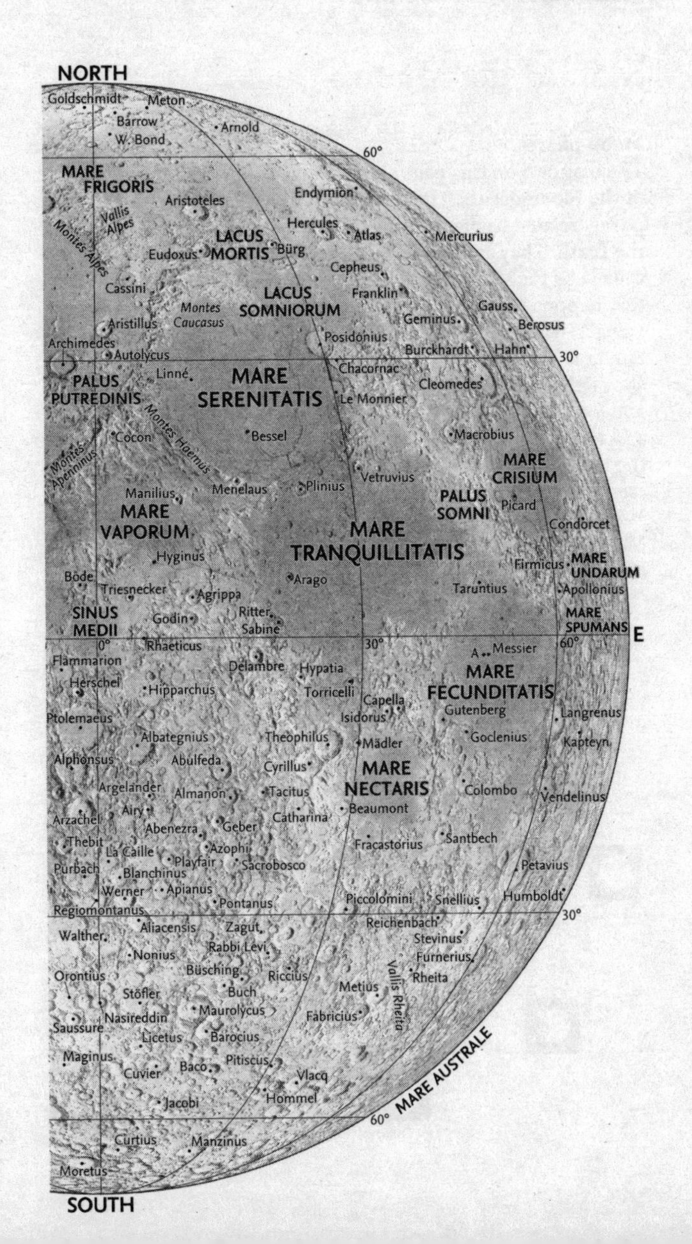
NORTH
Goldschmidt
Meton
Barrow
W. Bond
Arnold
60°
MARE FRIGORIS
Aristoteles
Endymion
Vallis Alpes
Montes Alpes
Hercules
Atlas
Mercurius
LACUS MORTIS
Eudoxus
Bürg
Cepheus
Cassini
LACUS SOMNIORUM
Franklin
Montes Caucasus
Gauss
Aristillus
Geminus
Berosus
Archimedes
Posidonius
Autolycus
Burckhardt
Hahn
30°
Linné
MARE SERENITATIS
Chacornac
PALUS PUTREDINIS
Cleomedes
Le Monnier
Montes Haemus
Bessel
Cocon
Macrobius
Montes Apenninus
MARE CRISIUM
Vetruvius
Manilius
Menelaus
Plinius
PALUS SOMNI
Picard
MARE VAPORUM
MARE TRANQUILLITATIS
Condorcet
Hyginus
Firmicus
MARE UNDARUM
Bode
Arago
Triesnecker
Agrippa
Taruntius
Apollonius
SINUS MEDII
Godin
Ritter
Sabine
MARE SPUMANS
E
0°
Rhaeticus
30°
60°
Messier
Flammarion
Delambre
Hypatia
MARE FECUNDITATIS
Herschel
Hipparchus
Torricelli
Capella
Ptolemaeus
Isidorus
Gutenberg
Langrenus
Albategnius
Theophilus
Mädler
Goclenius
Kapteyn
Alphonsus
Abulfeda
Cyrillus
MARE NECTARIS
Argelander
Almanon
Tacitus
Colombo
Vendelinus
Airy
Beaumont
Arzachel
Abenezra
Geber
Catharina
Thebit
Santbech
Fracastorius
La Caille
Azophi
Purbach
Playfair
Sacrobosco
Petavius
Blanchinus
Werner
Apianus
Pontanus
Piccolomini
Snellius
Humboldt
Regiomontanus
30°
Reichenbach
Walther
Aliacensis
Zagut
Stevinus
Rabbi Levi
Nonius
Furnerius
Orontius
Busching
Riccius
Rheita
Vallis Rheita
Stöfler
Buch
Metius
Nasireddin
Maurolycus
Fabricius
Saussure
Licetus
Barocius
MARE AUSTRALE
Maginus
Pitiscus
Cuvier
Baco
Vlacq
Jacobi
Hommel
60°
Curtius
Manzinus
Moretus
SOUTH

Moon phases

The diagram on this page shows how the different appearance of the Moon occurs. The changes in its apparent shape are purely related to the position of the Moon in its orbit around the Earth. They are not, as some people mistakenly believe, caused by the shadow of the Earth on the Moon. (The only time that happens is during a rare lunar eclipse, as described on pages 10–11.) Such events occur only at Full Moon, when the Sun and Moon are on opposite sides of the Earth. The diagram shows that New Moon is when the Moon lies between the Sun and Earth, and Full Moon when the Earth is between the Sun and the Moon. The age of the Moon (in days) is reckoned from the time of New Moon. After New Moon we have a waxing crescent until First Quarter, after which the Moon is described as waxing gibbous. After Full Moon we have a waning gibbous Moon until Last Quarter. Following that, we have a waning crescent until the next New Moon, when the sequence repeats.

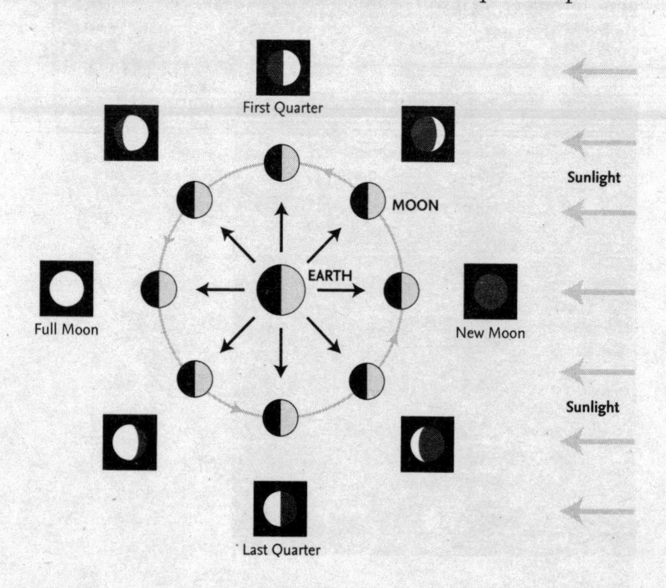

The Circumpolar Constellations

The northern circumpolar constellations

Learning the patterns of the stars, the constellations and asterisms is not particularly difficult. You need to start by identifying the various constellations that are circumpolar where you live. These are always above the horizon, so you can generally start at any time of the year. The charts on pages 30 and 32 show the northern and southern circumpolar constellations, respectively. The fine, dashed lines indicate the areas that are circumpolar at different latitudes.

The key constellation when learning the pattern of stars in the northern sky is ***Ursa Major***, in particular the seven stars forming the asterism known to many as the '***Plough***' or to people in North America as the '***Big Dipper***'. As the chart shows, this is just circumpolar for anyone at latitude 40°N, except for ***Alkaid*** (η Ursae Majoris), the last star in the 'tail'. Even so, the asterism of the Plough is low on the northern horizon between September and November, so it will be much easier to make out at other times of the year.

The position of the Big Dipper (the Plough), throughout the year, in relation to the northern horizon and Polaris, the Pole Star.

The northern circumpolar constellations.

The two stars ***Dubhe*** and ***Merak*** (α and β Ursae Majoris) are known as the 'Pointers', because they indicate the position of ***Polaris***, the Pole Star (α Ursae Minoris), at about a distance of five times their separation. Following this line takes you to the constellation of ***Ursa Minor***, the 'Little Bear' or 'Little Dipper', where Polaris is at the end of the 'tail' or 'handle'.

On the far side of the Pole is the constellation of ***Cassiopeia***, which is highly distinctive, with its five main stars forming the letter 'M' or 'W', depending on its orientation. Cassiopeia is circumpolar for observers at latitude 40°N or closer to the North Pole, although at times it is near the northern horizon and more difficult to see. (But at such times Ursa Major is clearly visible.) To find Cassiopeia from Ursa Major, start at ***Alioth*** (ε Ursae Majoris) and extend a line from that star to Polaris and beyond. It points to the central star of the five.

Moving anticlockwise from Cassiopeia, we come to ***Cepheus***, which has been likened to the gable end of a house, with its base in the Milky Way. The line from the Pointers to Polaris, if extended points to ***Errai*** (γ Cephei), at the top of the 'gable'. Continuing in the same direction, we come to ***Draco***, which wraps around Ursa Minor. The quadrilateral of stars that forms the 'head' of Draco is just circumpolar for observers at 40°N latitude, although it is brushing the horizon in January. On the opposite side of the sky to the head of Draco, the whole of the faint constellation of ***Camelopardalis*** is visible.

For observers slightly farther north, say at 50°N, additional constellations become circumpolar. The most important of these are ***Perseus***, not far from Cassiopeia, and most of which is visible and, farther round, the northern portion of ***Auriga***, with bright ***Capella*** (α Aurigae). On the other side of the sky is ***Deneb***, the brightest star in ***Cygnus***, although it is often close to the horizon, especially during the early night during the winter months. ***Vega*** (α Lyrae) another of the three stars that form the Summer Triangle is even farther south, often brushing the northern horizon, and only truly circumpolar and clearly seen at any time of the year for observers at 60°N.

Such far northern observers will also find that ***Castor*** (α Geminorum) is actually circumpolar, although at times it is extremely low on the horizon. The other bright star in ***Gemini***, ***Pollux*** (β Geminorum) is slightly farther south and cannot really be considered circumpolar.

The southern circumpolar constellations.

Eta Carinae (η Carinae) is one of the most massive and luminous stars known. It is estimated to have a mass between 120 and 150 times that of the Sun, and be between four and five million times as luminous.

The southern circumpolar constellations

Just as Ursa Major is the key constellation in the northern sky, so is ***Crux*** (the Southern Cross) an easily recognized feature of the southern circumpolar sky, although at times it may be brushing the horizon for observers at 30°S – roughly the latitude of Sydney in Australia. This is particularly true in the southern spring. Northeners, new to the southern sky, sometimes mistake the '***False Cross***', which consists of two stars from each of the constellations of ***Vela*** and ***Carina*** for the true Southern Cross. Crux itself is accompanied by the clearly visible dark cloud of the 'Coalsack' and also the two brightest stars in ***Centaurus: Rigil Kentaurus*** and ***Hadar*** (α and β Centauri, respectively). Together, the four stars of Crux and the two from Centaurus act as principal guides to the southern constellations.

Unfortunately, unlike the situation in the north, there is no star conveniently located at the South Celestial Pole (SCP), which lies in a relatively empty region of sky in the faint, triangular constellation of ***Octans***. Octans itself is perhaps best found by using the stars of ***Pavo*** as guides. A line from ***Peacock*** (α Pavonis) through the second brightest star (β) in that constellation, if

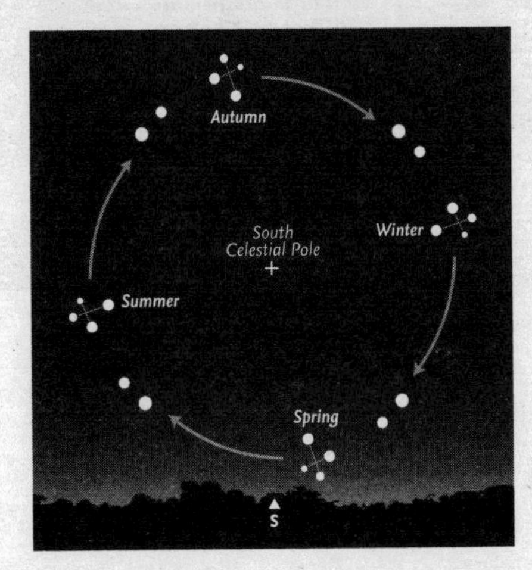

The position of Crux, the Southern Cross, throughout the year, in relation to the southern horizon. It also shows the position of the two brightest stars in Centaurus.

extended by about the same amount as the distance between the two stars, points to the 'base' of the triangle of Octans.

The main 'upright' of Crux, if extended and curving slightly to the right, does point in the approximate direction of the Pole, passing through ***Musca*** and the tip of ***Chamaeleon***. However, a better way is to start at Hadar (the star in the bright pair that is closest to Crux), turn at right-angles at Rigil Kentaurus, and following an imaginary line through the brightest star in the small constellation of ***Circinus*** and then right across the sky, brushing past the outlying star of ***Apus***, and the star (δ) at the apex of Octans itself.

Centaurus is a straggling constellation, with many stars well north of Rigil Kentaurus and Hadar, and some fainter ones that partially enclose Crux. Starting at Crux, and moving clockwise (in the same direction as the sky rotates), we come to the stars of ***Carina*** and ***Vela***, both originally part of the large, now obsolete constellation of Argo Navis. Past the False Cross, we come to ***Canopus*** (α Carinae), the second brightest star in the sky, which is just circumpolar for observers at 40°S, although occasionally, especially in June, very low over the northern horizon. Lying between Canopus and the SCP is the ***Large Magellanic Cloud*** (LMC), a satellite galaxy to our own. It lies across the boundaries of the constellations of ***Dorado*** and ***Mensa***.

Continuing round from Canopus we pass the constellations of Dorado, the small constellation of ***Reticulum*** and the undistinguished constellation of ***Horologium***, beyond which is ***Achernar*** (α Eridani) the brightest star in the long, winding constellation of ***Eridanus***, which actually starts far to the north, close to ***Rigel*** in ***Orion***. Between Achernar and the SCP lies the triangular constellation of ***Hydrus***, next to the constellation of ***Tucana*** which contains the ***Small Magellanic Cloud*** (SMC).

For observers farther south (at say, 50°S) there are the constellations of ***Phoenix***, followed by the roughly cross-shaped constellation of ***Grus*** and the rather faint ***Indus***. For observers at 40°S, the whole of the constellation of Pavo is visible, including its brightest star, Peacock. Farther round there is the constellation of ***Ara*** and, for observers farther north at 30°S, ***Triangulum Australe*** is fully visible.

Of the 88 constellations, the largest is ***Hydra*** with an area of 1303 square degrees. It covers more than 7 hours of right ascension (105° in the sky). It is more than 19 times as large as the smallest, **Crux**, which has an area of just 68 square degrees. ***Equuleus*** (87th), is not much larger with an area of 72 square degrees. ***Sagitta*** is the 86th, area 80 square degrees, and ***Circinus***, 85th, with an area of 93 square degrees. None of these four small constellation includes any objects of particular interest.

The Monthly Maps

How to use the monthly maps

The charts in this book are designed to be used more-or-less anywhere in the world. They are not suitable to be used at very high northern or southern latitudes (beyond 60°N or 60°S). That is slightly less than the latitudes of the Arctic and Antarctic Circles, beyond which there are approximately six months of daylight, followed by six months of darkness. The design may seem a little complicated, but these diagrams should make their usage clear. The main charts are given in pairs, one pair for each month: Looking North and Looking South.

Obviously, the region of the sky that is visible at any time entirely depends on one's location on Earth. You should imagine a rectangular 'window', 90° degrees high, that includes the sky from the horizon to the zenith. Think of moving this 'window' north or south over the charts, depending on your actual latitude. The base will be at your actual latitude on Earth. The other edge will be at your zenith. (You could make an actual 'window', by drawing a rectangle on a thin sheet of plastic, with the horizon and zenith lines 90° apart, and then use this on the charts.) The diagram on the next page shows this 'window' for the latitude of 50°N.

The scales on the right-hand and left-hand margins indicate the northern (or southern) horizon, for looking north (or south) respectively. The two diagrams on page 38 are drawn to indicate the horizon for the latitude of 40°N (the latitude of Philadelphia in the United States or Madrid in Spain); the second pair on page 39 show what would be visible 'looking north' and 'looking south' from latitude 30°S (the latitude of Durban in South Africa). If you are looking north (or south), once you get to the zenith, you can switch to the other chart, showing the view from the southern (or northern) horizon to the zenith.

To help you choose the correct latitude, there is a world map on pages 40–41.

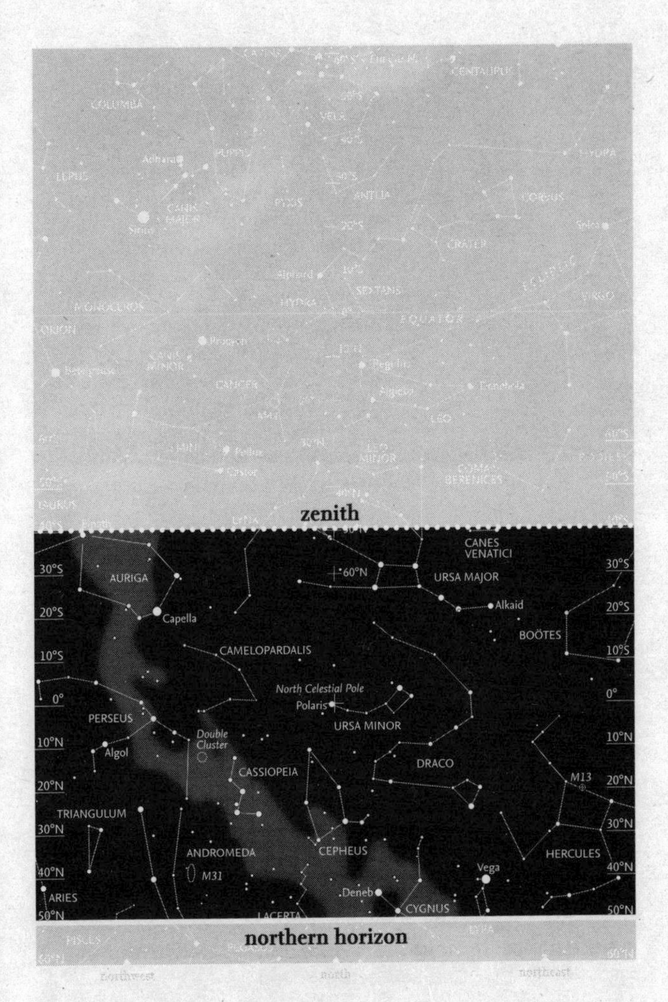

Horizon window, from the northern horizon (solid line at the bottom) to the zenith (the dotted line) for the latitude of 50°N.

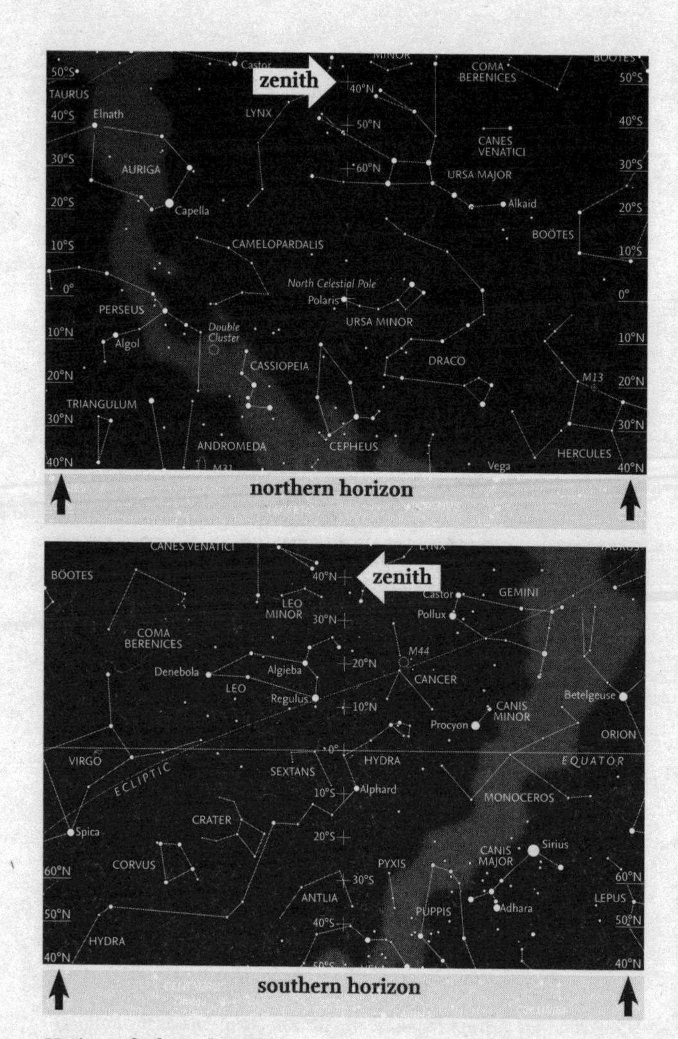

Horizons for latitude 40°N.

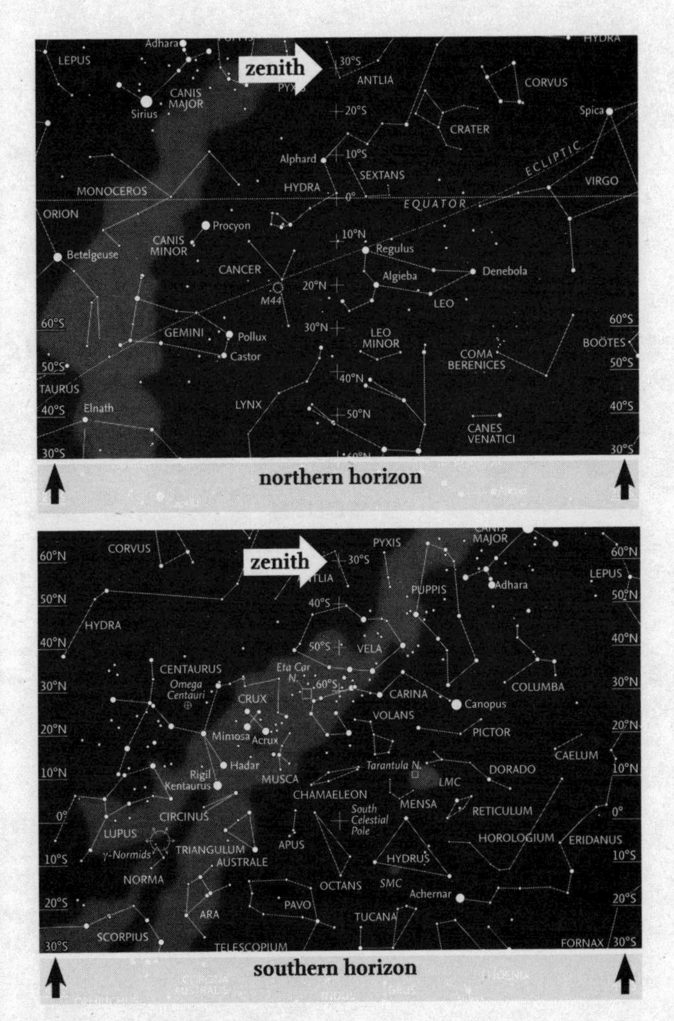

Horizons for latitude 30°S.

World Map

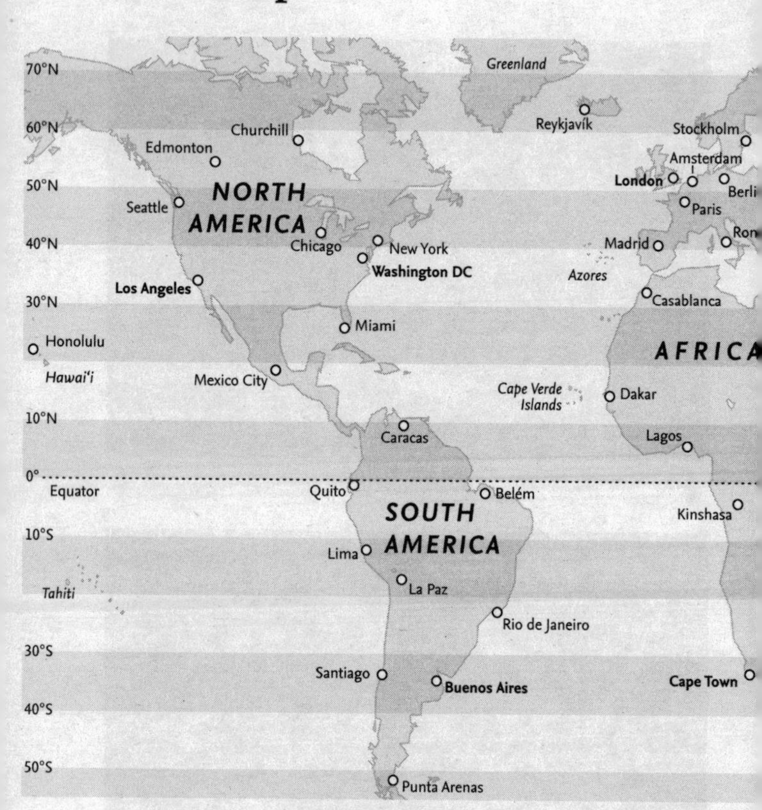
70°N
60°N
50°N
40°N
30°N
10°N
0°
Equator
10°S
30°S
40°S
50°S
Greenland
Reykjavík
Churchill
Edmonton
Stockholm
Amsterdam
London
Berli
Paris
Ron
NORTH AMERICA
Seattle
Chicago
New York
Washington DC
Madrid
Azores
Los Angeles
Casablanca
Miami
Honolulu
Hawai'i
AFRICA
Mexico City
Cape Verde Islands
Dakar
Caracas
Lagos
Quito
Belém
SOUTH AMERICA
Kinshasa
Lima
La Paz
Tahiti
Rio de Janeiro
Santiago
Buenos Aires
Cape Town
Punta Arenas

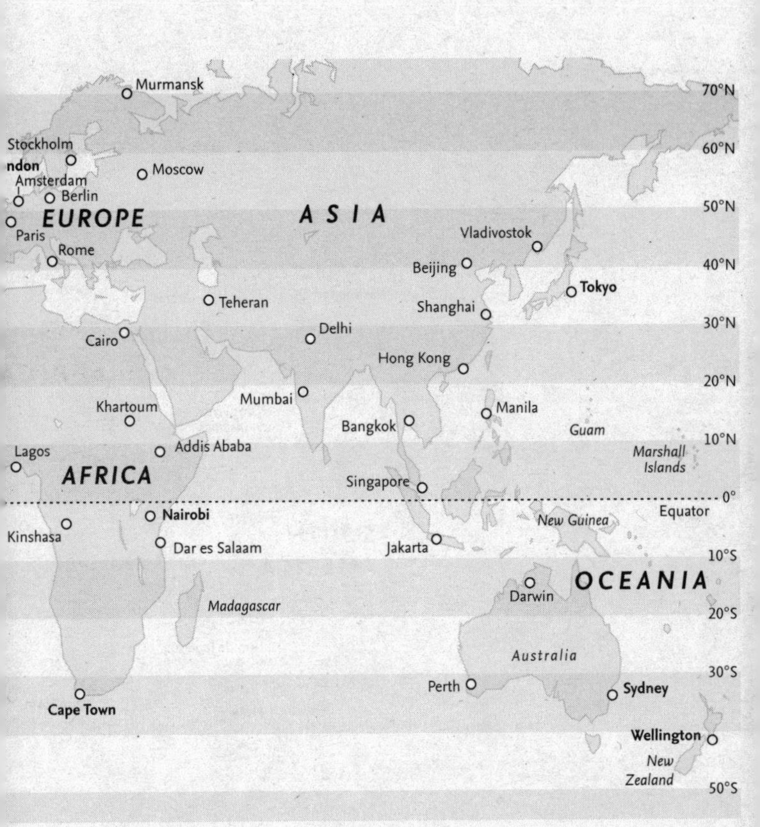
Murmansk
Stockholm
ndon
Amsterdam
Berlin
Moscow
EUROPE
ASIA
Paris
Rome
Vladivostok
Beijing
Tokyo
Teheran
Shanghai
Cairo
Delhi
Hong Kong
Mumbai
Khartoum
Manila
Bangkok
Guam
Lagos
Addis Ababa
Marshall Islands
AFRICA
Singapore
Nairobi
New Guinea
Equator
Kinshasa
Dar es Salaam
Jakarta
OCEANIA
Darwin
Madagascar
Australia
Perth
Sydney
Cape Town
Wellington
New Zealand
70°N
60°N
50°N
40°N
30°N
20°N
10°N
0°
10°S
20°S
30°S
50°S

January

January – Introduction

The Earth

The Earth reaches perihelion (the closest point to the Sun in its yearly orbit) on January 4 at 06.55 Universal Time, when its distance from the Sun is 0.983336540 Astronomical Units (AU) or 147,104,983.044 km.

There are no major astronomical events during January 2022, but it is the month in which the peak of the annual ***Quadrantid*** meteor shower occurs (see page 56). This is a shower that is visible from the northern hemisphere only, with an extremely sharp peak in January. Unfortunately, because of the cold conditions and weather at that time of year, the shower tends to be poorly observed.

Occultations

There are no occultations in 2022 of the five bright stars near the ecliptic (***Aldebaran***, ***Antares***, ***Pollux***, ***Regulus*** and ***Spica***.) There are two occultations of Venus (on May 27 and October 25), but these are not readily observable. There are three occultations of Mars during the year (on June 22, July 21, December 8). Only the last of these may be observable, with difficulty. These planetary occultations will be described in more detail in the appropriate months.

The planets

The inner planets, ***Mercury*** and ***Venus***, are close to the Sun in January and largely lost in twilight. ***Mars*** at mag. 1.4, is initially in ***Ophiuchus***, but moves into ***Sagittarius*** and is difficult to see in the morning twilight. ***Jupiter*** is visible in the evening sky in ***Aquarius*** at mag. -2.1. ***Saturn*** is in ***Capricornus*** and too close to the Sun to be easily visible. ***Uranus*** is mag. 5.7 in ***Aries***, resuming direct motion (moving eastwards) on January 22, and ***Neptune*** is mag. 7.9 in ***Aquarius***. On January 13, the minor planet (7) ***Iris*** is at opposition in ***Gemini*** at mag. 7.6 (see the charts on the next page).

The ***Huygens*** probe landed on Saturn's satellite, Titan, on 14 January 2005.

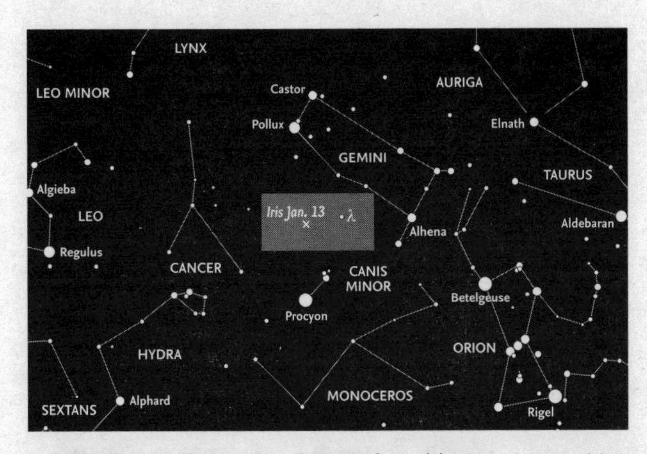

A finder chart for the position of minor planet (7) Iris, at its opposition. The grey area is shown in more detail on map below.

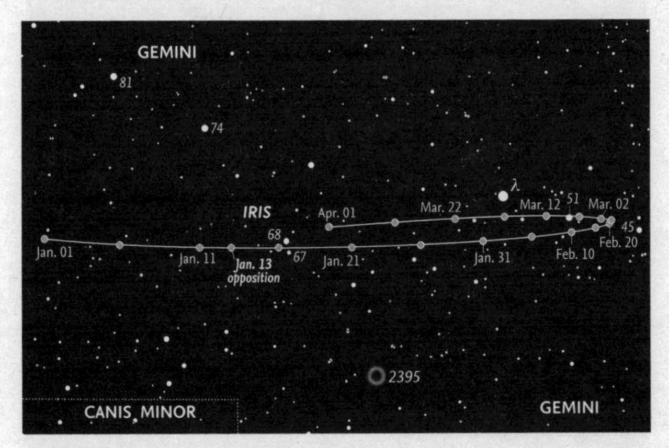

The path of the minor planet (7) Iris around its opposition on January 13 (mag. 8.5). Background stars are shown down to magnitude 9.0.

Sunrise and Sunset

City	Date	Sunrise	Sunset
Buenos Aires, Argentina			
	Jan. 01	08:44	23:10
	Jan. 31	09:13	23:01
Cape Town, South Africa			
	Jan. 01	03:38	18:01
	Jan. 31	04:06	17:52
London, UK			
	Jan. 01	08:07	16:02
	Jan. 31	07:41	16:48
Los Angeles, USA			
	Jan. 01	14:59	00:54
	Jan. 31	14:51	01:22
Nairobi, Kenya			
	Jan. 01	03:30	15:42
	Jan. 31	03:41	15:51
Sydney, Australia			
	Jan. 01	18:48	09:09
	Jan. 31	19:16	09:01
Tokyo, Japan			
	Jan. 01	21:51	07:38
	Jan. 31	21:41	08:07
Washington, DC, USA			
	Jan. 01	12:27	21:57
	Jan. 31	12:15	22:29
Wellington, New Zealand			
	Jan. 01	16:52	07:57
	Jan. 31	17:26	07:43

NB: the times given are in Universal Time (UT)

J

The Moon's Phases and Ages

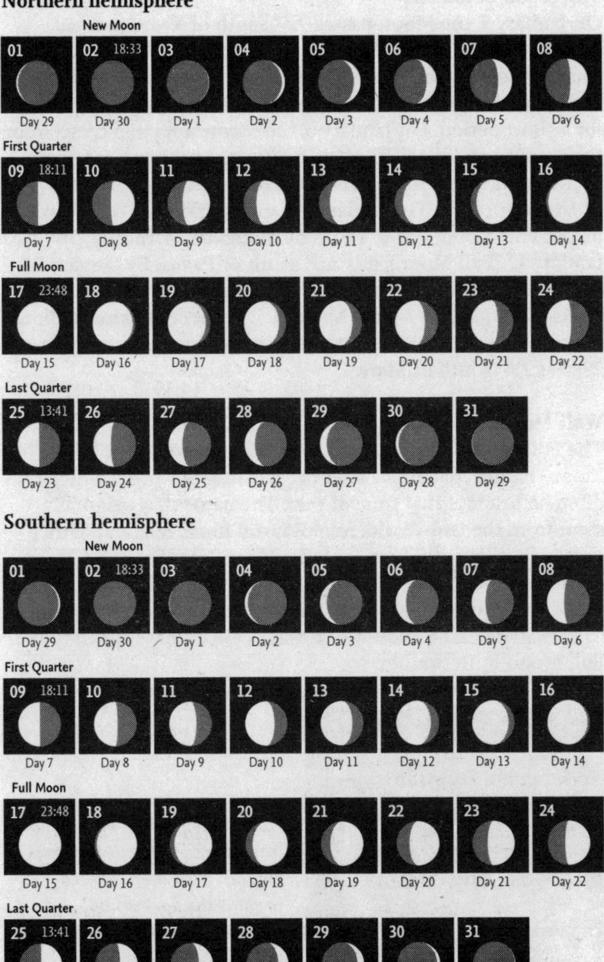

The Moon

The Moon in January

On January 3, the Moon passes 7.5° south of ***Venus*** and on January 4, 3.1° south of ***Mercury***, followed, later that day, 4.2° south of ***Saturn***. The first of these events is very close to the Sun and difficult to see, but it may be possible to glimpse the others for a short period. On January 6, the Moon, a waxing crescent in the evening sky, is 4.5°south of ***Jupiter*** in ***Aquarius***. On January 11, two days after First Quarter, the Moon passes 1.5° south of ***Uranus*** in ***Aries***. On January 14, waxing gibbous, three days before Full Moon, it is 6° north of ***Aldebaran*** in ***Taurus***. On January 17 (Full Moon), it is 2.6° south of ***Pollux***. By January 20, it is in Leo, passing 4.0° north of ***Regulus*** and between it and ***Algieba***. On January 24, the Moon is 5.5° north of ***Spica***, and on January 27 it passes 3.7° north of ***Antares***. One day later (January 28) it is 2.4° south of ***Mars***.

Wolf Moon

The Full Moon in January is sometimes known as the 'Wolf Moon', named, of course, after the howling of wolves, which may often be heard at that time of year. The name may originally stem from the Old-World, Anglo-Saxon lunar calendar. Other names for this Full Moon include: Moon After Yule, Old Moon, Ice Moon, and Snow Moon. Among the Algonquin tribes, the name was 'squochee kesos', meaning 'the Sun has not strength to thaw'. The name 'Wolf Moon' was occasionally applied to the Full Moon in December.

In this photograph, the narrow lunar crescent (about two days old) has been over-exposed to show the Earthshine illuminating on the other portion of the Moon, where the dark maria are faintly visible.

Names of the Full Moon

The Native American tribes had a set of names for the Full Moon, depending on the time of year. The actual names varied between different tribes, so there may be more than one name used for a particular Full Moon. The interval between successive Full Moons (or between any other specific phases of the Moon) is known as the synodic month, and is, on average, 29.53 days, so the names have come to be associated with modern calendar months. They are:

- January: 'Wolf Moon'
- February: 'Snow Moon', 'Hunger Moon', 'Storm Moon'
- March: 'Worm Moon', 'Crow Moon', 'Sap Moon', 'Lenten Moon'
- April: 'Seed Moon', 'Pink Moon', 'Sprouting Grass Moon', 'Egg Moon', 'Fish Moon'
- May: 'Milk Moon', 'Flower Moon', 'Corn Planting Moon'
- June: 'Mead Moon', 'Strawberry Moon', 'Rose Moon', 'Thunder Moon'
- July: 'Hay Moon', 'Buck Moon', 'Elk Moon', 'Thunder Moon'
- August: 'Corn Moon', 'Sturgeon Moon', 'Red Moon', 'Green Corn Moon', 'Grain Moon'
- September: 'Harvest Moon', 'Full Corn Moon'
- October: 'Hunter's Moon', 'Blood Moon'/'Sanguine Moon'
- November: 'Beaver Moon', 'Frosty Moon'
- December: 'Oak Moon', 'Cold Moon', 'Long Night's Moon'

The only two names commonly used in Europe were 'Harvest Moon' and 'Hunter's Moon'. On rare occasions, particularly in religious contexts, the term 'Lenten Moon' was used for the Full Moon in March. The other terms, originating in North America, have been adopted increasingly by the media in recent years.

Calendar for January

01–12		Quadrantid meteor shower
01	22:55	Moon at perigee (358,033 km)
02	18:33	New Moon
03–04		Quadrantid meteor shower maximum
03	08:09	Venus 7.5°N of Moon
04	01:21	Mercury 3.1°N of Moon
04	06:55	Earth at perihelion (0.983337 AU = 147,104,983 km)
04	16:47	Saturn 4.2°N of Moon
06	00:11	Jupiter 4.5°N of Moon
07	09:46	Neptune 4.1°N of Moon
07	11:04	Mercury at greatest elongation (19.2°E, mag. -0.6)
09	00:48	Venus at inferior conjunction
09	18:11	First Quarter
11	11:27	Uranus 1.5°N of Moon
13	20:35	Minor planet (7) Iris at opposition (mag. 7.6)
14	01:35	Aldebaran 6°S of Moon
14	09:26	Moon at apogee = 405,805 km
17	16:13	Pollux 2.6°N of Moon
17	23:48	Full Moon
20	11:00	Regulus 4.9°S of Moon
23	10:28	Mercury at inferior conjunction
24	13:59	Spica 5.5°S of Moon
25	13:41	Last Quarter
27	23:28	Antares 3.7°S of Moon
29	15:03	Mars 2.4°N of Moon
30	01:50	Venus 10.2°N of Moon
30	07:11	Moon at perigee = 362,252 km
31–Feb.20		α-Centaurid meteor shower
31	00:18	Mercury 7.6°N of Moon

January 3–5 • *After sunset, the narrow crescent Moon passes Mercury, Saturn and Jupiter (as seen from central USA).*

January 17 • *The Full Moon lines up with Castor and Pollux. Procyon is close to the eastern horizon (as seen from London).*

January 20/21 • *Around midnight the Moon is between Regulus and Algieba (γ Leo). As seen from Sydney.*

January 28–29 • *The waning crescent Moon passes Sabik and Mars. Venus is farther east (as seen from central USA).*

LOOKING NORTH

Early January 23:00 — Mid January 22:00 — Late January 21:00

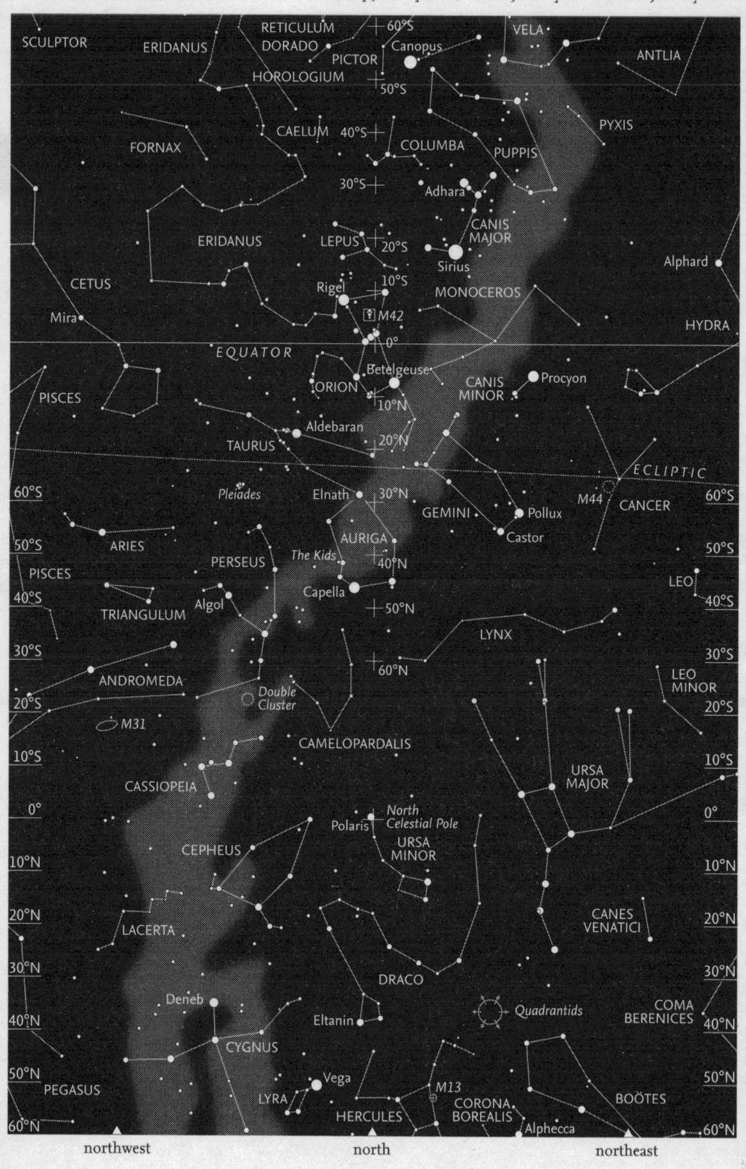

January – Looking North

For most northern observers, all the important northern circumpolar constellations (see pages 29–31) of ***Ursa Major***, ***Ursa Minor*** (with ***Polaris***, the Pole Star), ***Cassiopeia***, ***Draco*** and ***Cepheus*** will be visible. Polaris is, of course, the star about which the sky appears to rotate, even though it is not precisely at the North Celestial Pole. Ursa Major stands more-or-less vertically above the horizon in the northeast. Opposite it in the northwest of the sky is the 'W' of Cassiopeia. (Looking more like an 'M' at this time of the year.) For most observers, ***Capella*** (α Aurigae) is high overhead, but only those at high latitudes will find it easy to see the quadrilateral of stars that marks the head of Draco, brilliant ***Deneb*** (α Cygni) or the even brighter ***Vega*** (α Lyrae), yet farther south. Deneb and ***Eltanin*** (γ Draconis), the brightest star in the 'Head' of Draco, are skimming the horizon for observers at 40°N.

The ***Apollo 14*** manned lunar mission was launched on 31 January 1971.

For observers between about 30 and 50°N, the constellation of ***Auriga*** is near the zenith (and thus difficult to observe). This important constellation contains bright ***Capella*** (α Aurigae) and farther to its west lies the constellation of ***Perseus***, with ***Algol***, the famous variable star.

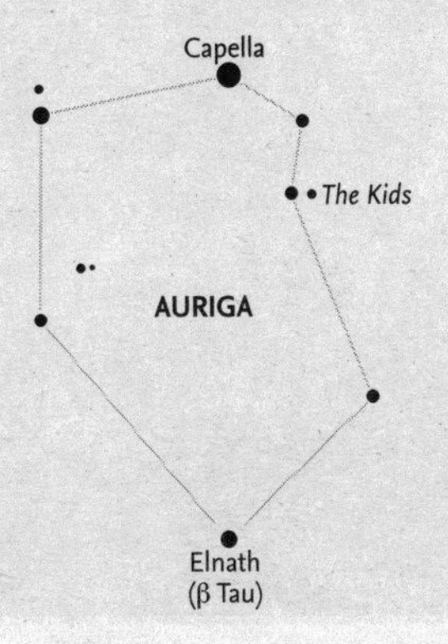

The constellation of Auriga, with brilliant Capella which, although appearing as a single star, is actually a quadruple system, consisting of a pair of yellow giant stars, gravitationally bound to a more distant pair of red dwarfs. Elnath, near the bottom, actually belongs to the constellation of Taurus.

LOOKING SOUTH

Early January 23:00 — Mid January 22:00 — Late January 21:00

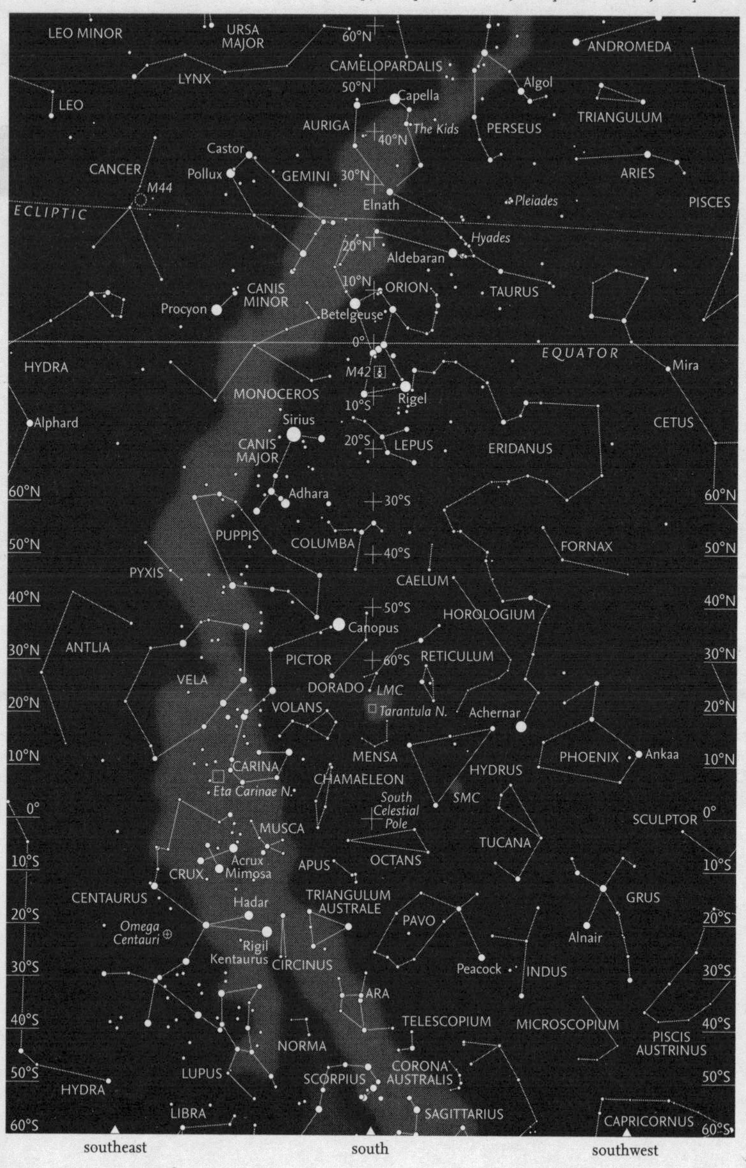

January – Looking South

J

The southern sky is dominated by ***Orion***, visible from nearly everywhere in the world and prominent during the northern winter months. For observers near the equator it is, of course, high above near the zenith. Orion is highly distinctive, with a line of three stars that form the 'Belt'. To most observers, the bright star ***Betelgeuse*** (α Orionis), shows a reddish tinge, in contrast to the brilliant bluish-white ***Rigel*** (β Orionis). The three stars of the belt lie directly south of the celestial equator. A vertical line of three 'stars' forms the 'Sword' that hangs south of the Belt. With good viewing, the central 'star' appears as a hazy spot, even to the naked eye, and is actually the ***Orion Nebula*** (M42). Binoculars reveal the four stars of the Trapezium, which illuminate the nebula.

Orion's Belt points up to the northwest towards ***Taurus*** (the Bull) and orange-tinted ***Aldebaran*** (α Tauri). Close to Aldebaran, there is a conspicuous 'V' of stars, called the ***Hyades*** cluster. (Despite appearances, Aldebaran is not part of the cluster.) Farther along, the same line from Orion passes below a bright cluster of stars, the ***Pleiades***, or Seven Sisters. Even the smallest pair of binoculars reveals this as a beautiful group of bluish-white stars. The two most conspicuous of the other stars in Taurus lie directly north of Orion, and form an elongated triangle with Aldebaran. The northernmost, ***Elnath*** (β Tauri), was once considered to be part of the constellation of ***Auriga***.

Slightly to the west of ***Capella*** lies a small triangle of fainter stars, known as '***The Kids***'. (Ancient mythological representations of Auriga show him carrying two young goats.) Together with Elnath, the body of Auriga forms a large pentagon on the sky, with The Kids lying on the western side (see page 53).

Running south from Orion is the long constellation of ***Eridanus*** (the River), which begins near Rigel in Orion and runs far south to end at ***Achernar*** (α Eridani). To the south of Orion is the constellation of ***Canis Major*** and several other constellations, including the oddly shaped ***Carina***. The line of Orion's Belt also points southeast in the general direction of ***Sirius*** (α Canis Majoris), the brightest star. Almost due south of Sirius lies ***Canopus*** (α Carinae), the second brightest star in the sky.

Meteors

Quadrantids

The Quadrantid meteor shower (named after the obsolete constellation of ***Quadrans Muralis***) is active from the end of December (December 28) to January 12. It is one of the strongest showers of the year, with rates similar to those of the more prominent showers of the Perseids (in August) and Geminids (in December). There is, however, a very sharp peak, lasting no more than about 6 hours, and this next occurs on 3–4 January 2022. This sharp peak may be easily missed if there is bad weather or moonlight. (The Moon is a waning crescent in 2022, so conditions are very favourable.) The overall brightness of the meteors is low, but it does frequently produce bright fireballs.

The radiant lies roughly half-way between θ Boötis and τ Herculis, and may also be envisaged as lying between ***Alkaid*** (η UMa) – the end of the 'tail' of ***Ursa Major*** (the end of the handle of the Big Dipper) – and the head of ***Draco***. Because of the radiant's location, the shower is not well seen from the southern hemisphere, although occasional meteors may be seen from as far south as 50°S.

The parent body has been tentatively identified as the minor planet 2003 EH1, which may be linked to comet C/1490 Y1, observed by Far-Eastern astronomers (Chinese, Japanese, and Korean) some 530 years ago.

A brilliant, very late Quadrantid fireball, photographed by Denis Buczynski from Portmahomack, Ross-shire, Scotland, on 15 January 2018, at 23:44 UT.

Auzout

Adrien Auzout was a French astronomer, born in Rouen on 28 January 1622 and died on 23 May 1691. Among his achievements, he is noted for being one of the founders of the Paris Observatory, his observations of comets and his arguments that cometary orbits were elliptical or parabolic. (This was long before, in 1705, Edmond Halley used his September 1682 observations and Isaac Newton's theories to suggest that the comet, subsequently named after him, followed an elliptical orbit and would return in 1758.) Comets had only been shown to be astronomical in nature, rather than meteorological, after observations by the famous astronomer, Tycho Brahe, of the Comet of 1577. A lunar crater, just south of Mare Crisium, is named after Auzout.

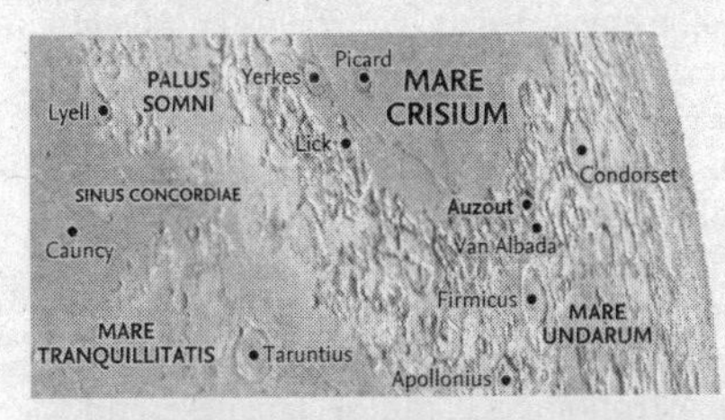

The location of the crater Auzout, just south of Mare Crisium.

The crater Auzout is the uppermost (to the north) in this image, which also shows the adjoining crater Van Albada.

On 23 January 2003, communications were lost with the ***Pioneer 10*** spaceprobe when it was at a distance of 1,200 million kilometers (80 AU) from Earth.

February

February – Introduction

February is a quiet month for astronomical events. For northern observers, there are no major meteor streams, but southern observers are able to observe two showers. The most important of these are the ***Centaurids***, which actually begin on January 31, and consist of two separate streams: the ***α-Centaurids*** and the ***β-Centaurids***, with radiants lying near those two stars (***Rigil Kentaurus*** and ***Hadar***, respectively). Both branches of this shower reach a low maximum, with an hourly rate of about 5–6 meteors per hour, on February 8. That day, the Moon is at Day 8 of the lunation, at First Quarter, so observing conditions are not particularly favourable.

The second southern shower, the ***γ-Normids***, begins to be active around February 25 and continues into March, reaching its weak (but very sharp) maximum on March 14–15. Unfortunately, its meteors are difficult to differentiate from sporadics, so are likely to be identified by dedicated meteor observers only. Observing conditions at maximum are very difficult in 2022, because of moonlight. The Moon is waxing gibbous, nearing Full (on March 16).

Three spaceprobes arrived in martian orbit in February 2021. The Emirates martian atmospheric-studies mission, ***Hope***, on February 9, and the Chinese ***Tianwen-1*** (orbiter, lander and rover) on February 10. The NASA ***Perseverance*** rover was successfully landed in Jezero crater on February 18.

The planets

Mercury is at greatest elongation 26.3° west of the Sun in the morning sky on February 16. ***Venus*** is in ***Sagittarius***, visible in the morning sky and 7° north of ***Mars*** on February 13. Mars is also in Sagittarius, but becomes very low towards the end of the month and difficult to see. ***Jupiter*** and ***Saturn*** are too close to the Sun to be visible. Saturn is at superior conjunction, on the far side of the Sun, on February 4. ***Uranus*** remains in ***Aries*** at mag. 5.8, and ***Neptune*** is in ***Aquarius*** at mag. 7.9 to 8.0.

The constellation of Orion dominates the evening sky during the first months of the year, and is a useful starting point for recognizing other constellations in the southern sky.

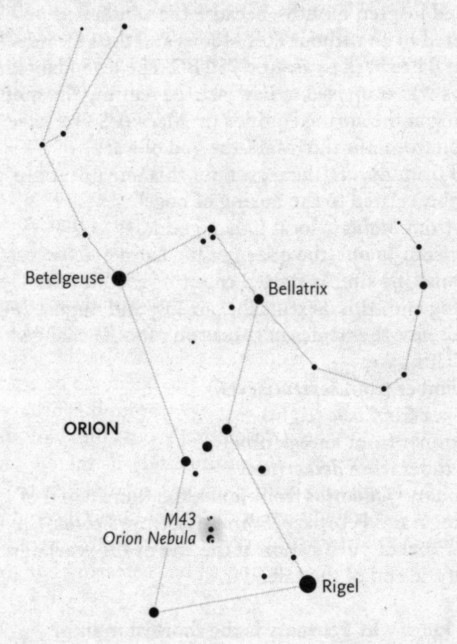

Did you know?
Why is February the shortest month? And why does it have such an odd number of days? The answers to these questions are surprisingly complicated. They involve the lunar calendar, ancient Romans, priests, Julius Caesar, and the way politicians tinkered with the calendar. A fairly comprehensive description of how these oddities came about is given on pages 62–63.

February, the Shortest Month

The names of the months

Several of our modern names for the months derive from the original Roman calendar. This 'calendar of Romulus' contained just ten months, because the winter was considered to be without Full Moons and thus months. The calendar dates back to around 750 BC. The legendary king Romulus was supposed to have started naming the months, beginning at the spring equinox (in March). So we have:

- March from ***Martius*** (Mars, the god of war)
- April from ***Aprilis*** (the reason for this one uncertain, and possibly related to the raising of hogs)
- May from ***Maius*** (a local Italian goddess)
- June from ***Junius*** (the queen of the Latin gods).

After Junius, he simply started counting the months beginning Quintilis, Sextilis for our July and August. (For details of how the names of these two months changed, see below.) This gave:

- September from ***septem*** (seven)
- October from ***octo*** (eight)
- November from ***novem*** (nine)
- December from ***decem*** (ten)

Two, January (***Januarius*** from ***Janus***, the Roman god of beginnings) and February (***Februarius***, from ***Februa***, the Roman ritual of purification at the end of the year) were later added at the end of the calendar.

Do you know why February is the shortest month?

February now has either 28 or 29 days. The reason is because originally the calendar was a lunar one, based on the Moon. A fixed number of lunar months does not agree with a solar year. The time from Full Moon to Full Moon (or between any other pair of phases), known to astronomers as the synodic month, is slightly more than 29.5 solar days. The best fit for a solar year is 365 days, although we occasionally add a day to keep the calendar in line with the Sun.

The trouble with February lies with the Romans. Originally, they used a lunar calendar, and the winter was considered to be without months. Then, about 713 BC, two months, January and February were added to the end of the year. (The year was supposed to begin at the spring equinox in March.) To keep the lunar calendar in step with the year, what are called intercalary months were sometimes inserted.

However, the calendar was initially under the control of priests and, as might be expected, they fiddled things to their advantage. The calendar was used to determine when people could carry out business, and when ceremonies – very important in ancient Rome – could occur. The priests added intercalary months whenever they thought fit; for example, when more taxes were required. There was utter chaos and the calendar never agreed with the solar year.

Eventually, Julius Caesar became 'dictator for life'. He instigated a calendar reform, and his calendar subsequently became known as the ***Julian calendar***. In this, months alternated between 30 and 31 days. This was slightly too long (6 × 30 + 6 × 31 = 366 days), so one day was removed from the last month of the year: February, giving it 29 days. An additional day was returned to February every four years (the leap year), to keep things in step with the Sun. One month, ***Quintilis***, was named ***Julius*** (the month of Caesar's birth).

But then the priests messed things up again. They started counting leap years as occurring every three years. So chaos ensued again. The error was corrected by the emperor Augustus and by AD 8 the matter had been solved and the months and the Sun were in agreement. But then the Senate decided to rename one month in honour of Augustus – so the month of ***Sextilis*** became our August. Unfortunately, under Caesar's scheme that month had just 30 days, whereas that honouring Caesar himself (our July) had 31 days. Obviously Augustus had to have the same number of days, so they pinched one from poor February, leaving it with 28 days, except in leap years. At the same time, to avoid having three months with 31 days in succession (July, August and September) they also tinkered with the lengths of the months after August, which is why September and November now have 30 days and October and December 31.

Sunrise and Sunset

City	Date	Sunrise	Sunset
Buenos Aires, Argentina			
	Feb. 01	09:14	23:00
	Feb. 28	09:40	22:31
Cape Town, South Africa			
	Feb. 01	04:07	17:51
	Feb. 28	04:33	17:24
London, UK			
	Feb. 01	07:40	16:50
	Feb. 28	06:49	17:39
Los Angeles, USA			
	Feb. 01	14:50	01:23
	Feb. 28	14:23	01:48
Nairobi, Kenya			
	Feb. 01	03:41	15:51
	Feb. 28	03:41	15:49
Sydney, Australia			
	Feb. 01	19:17	09:01
	Feb. 28	19:43	08:33
Tokyo, Japan			
	Feb. 01	21:41	08:08
	Feb. 28	21:12	08:35
Washington, DC, USA			
	Feb. 01	12:14	22:30
	Feb. 28	11:42	23:00
Wellington, New Zealand			
	Feb. 01	17:28	07:42
	Feb. 28	18:01	07:06

NB: the times given are in Universal Time (UT)

The Moon's Phases and Ages

Northern hemisphere

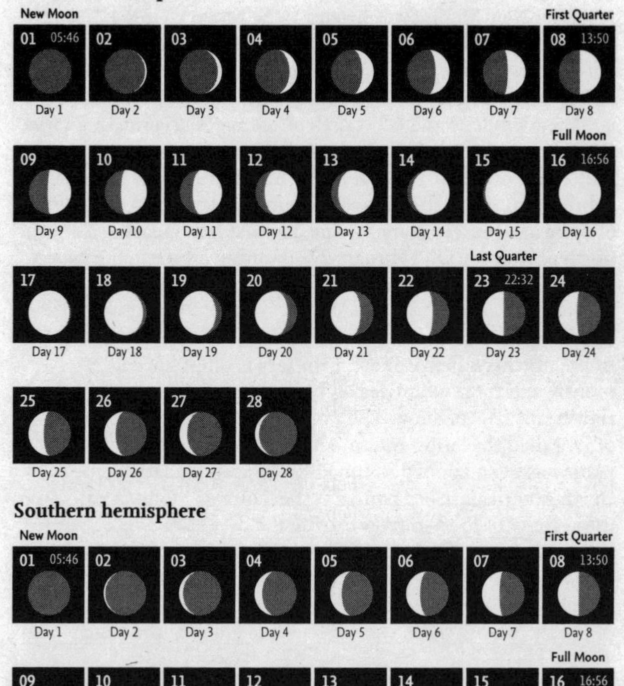

Southern hemisphere

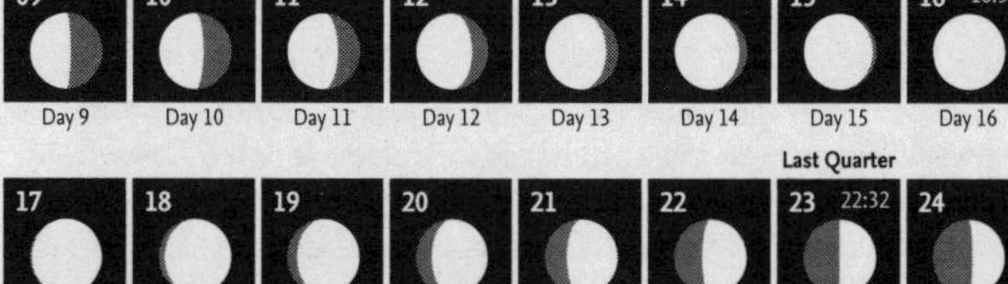

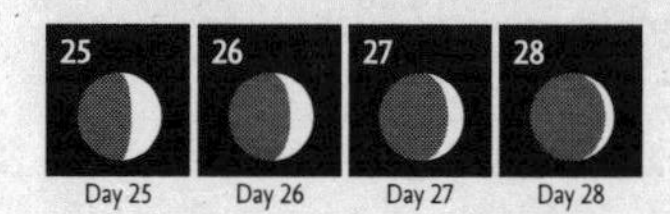

The Moon

The Moon in February

Just after New Moon on February 1, the Moon passes 4.2° south of ***Saturn***. The next day it is a similar distance south of ***Jupiter***, but very low on the western horizon. On February 7, the Moon is 1.3° south of ***Uranus*** in ***Aries***. Three days later, on February 10, it is waxing gibbous and 6.7° north of orange ***Aldebaran*** in ***Taurus***. On February 13 it is 2.6° south of ***Pollux*** in ***Gemini*** and just after Full Moon on February 16 it is 4.8° north of ***Regulus*** in ***Leo***. On February 20, when the Moon is waning gibbous, it is 5.3° north of ***Spica*** and, on February 24, one day past Last Quarter, 3.4° north of ***Antares***. On February 27, a waning crescent, the Moon is south of ***Venus*** and ***Mars***, and low in the morning twilight.

Snow Moon

In the northern hemisphere, February is often the coldest month, and most countries on both sides of the Atlantic see significant falls of snow. The Full Moon of February is thus often called the 'Snow Moon', although just occasionally that name has been applied to the Full Moon in January. Some North American tribes named it the 'Hunger Moon' because of the scarcity of food sources during the depths of winter, while other names are 'Storm Moon' and 'Chaste Moon', although the last name is more commonly applied to the Full Moon in March. To the Arapaho of the Great Plains, the Full Moon was called the Moon 'when snow blows like grain in the wind'.

Black Moon

Because it never has more than 29 days, and the synodic month (between any pair of phases, such as from New Moon to New Moon) is slightly more than half a day longer, sometimes there is no Full Moon in February. This occurs about every 19 years. This is one of the definitions of the term '***Black Moon***', although the same term is sometimes applied to a New Moon that does not occur in a particular season, as reckoned from the equinoxes or solstices.

The Winchcombe meteorite

At 21:54 on the evening of 28 February 2021, a bright fireball (magnitude -6 to -7) was observed over Gloucestershire. The trajectory was captured by various cameras, including those of the UK Fireball Alliance, together with videos from various doorbell cameras (of varying quality and usefulness). A reconstruction of its trajectory suggested that fragments might be recovered. The next day, several meteorites were found in Winchcombe and in a field in nearby Woodmancote in Gloucestershire. The largest meteorite, found in a driveway at Winchcombe, was a 300-g meteorite and appears to belong to the rare class of CM carbonaceous chondrites. These are extremely primitive meteorites, believed to resemble the material from which the planets were formed, some 4500 million years ago. Carbonaceous chondrite meteorites are described in more detail on pages 74–75. Because of the rapidity with which the fragment was recovered, it appears to be in an ideal, unaltered state, unaffected by terrestrial conditions and, in that respect, comparable with the material collected by the Japanese *Hayabusa* probe from the minor planet (162173) Ryugu and returned to Earth.

The main Winchcombe meteorite, recovered from a driveway in the market town.

Calendar for February

01	05:46	New Moon
01	08:57	Saturn 4.2°N of Moon
02	21:10	Jupiter 4.3°N of Moon
03	21:13	Neptune 3.9°N of Moon
04	19:05	Saturn at superior conjunction
07	19:39	Uranus 1.2°N of Moon
08		α-Centaurid meteor shower maximum
08	13:50	First Quarter
10	08:53	Aldebaran 6.7°S of Moon
11	02:37	Moon at apogee = 404,897 km
13	01:00 *	Mars 7.0°S of Venus
13	23:28	Pollux 2.6°N of Moon
16	16:56	Full Moon
16	17:46	Regulus 4.8°S of Moon
16	21:07	Mercury at greatest elongation (26.3°W, mag. -0.0)
20	19:34	Spica 5.3°S of Moon
23	22:32	Last Quarter
24	05:50	Antares 3.4°S of Moon
25–Mar.28		γ-Normid meteor shower
26	22:25	Moon at perigee = 367,789 km
27	06:28	Venus 8.7°N of Moon
27	08:59	Mars 3.5°N of Moon
28	20:05	Mercury 3.7°N of Moon
28	23:45	Saturn 4.3°N of Moon

* *These objects are close together for an extended period around this time.*

F

February 13 • *Venus, Mars and Mercury in the early morning. Nunki (σ Sgr) is close to Mars (as seen from central USA).*

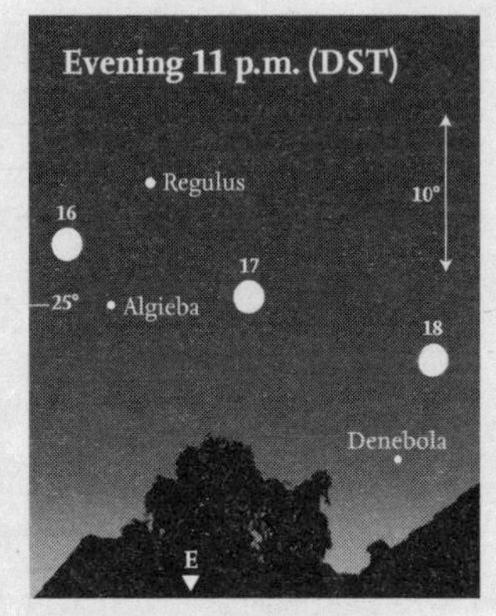

February 16–18 • *The Moon passes through the constellation of Leo, the Lion (as seen from Sydney).*

February 27 • *The crescent Moon with Venus and Mars. Nunki is farther towards the south (as seen from central USA).*

February 27–28 • *The waning crescent Moon passes Nunki, Venus and Mars (as seen from Sydney).*

Early February 23:00 — Mid February 22:00 — Late February 21:00

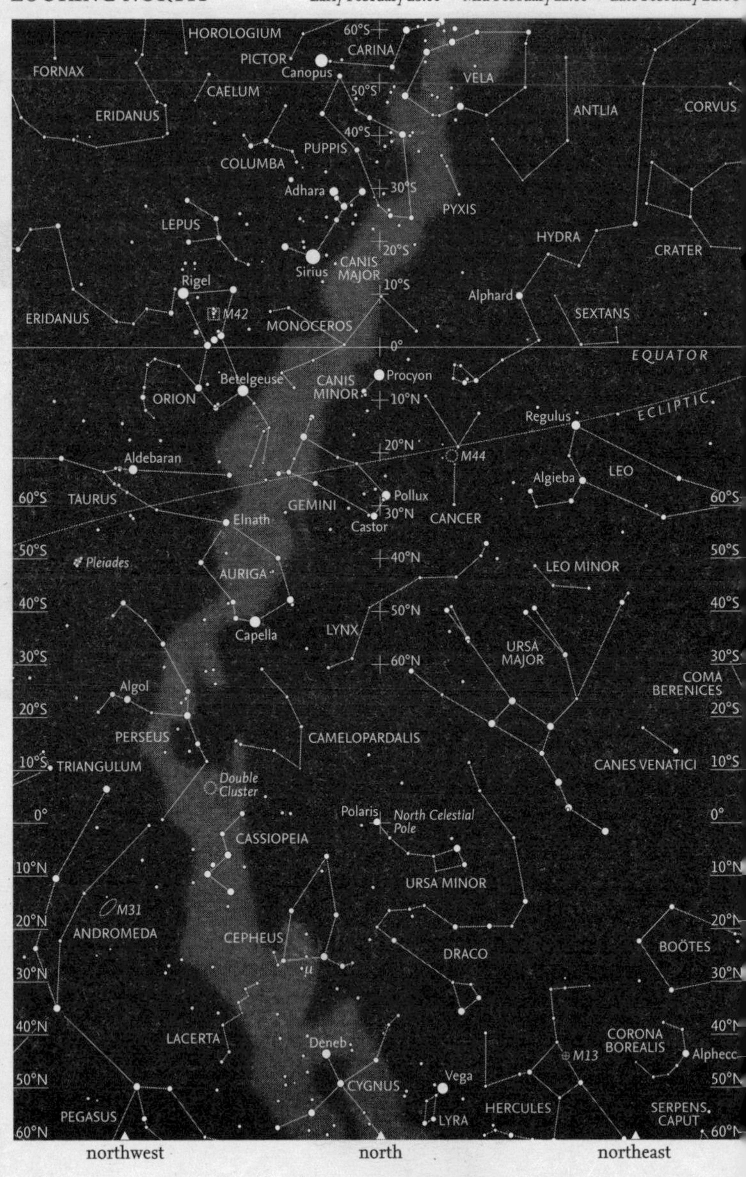

February – Looking North

The months of January and February are probably the best time for seeing the section of the Milky Way that runs in the northern and western sky from ***Orion*** and ***Gemini*** right through ***Auriga***, ***Perseus*** and ***Cassiopeia***, towards ***Cygnus***, low in the north. Although not as readily visible as the denser star clouds of the summer Milky Way, on a clear night so many stars may be seen that even a distinctive constellation such as Cassiopeia, which lies across the Milky Way, is not immediately obvious.

The 'base' of the constellation of ***Cepheus*** lies on the edge of the stars of the Milky Way, but the red supergiant star ***Mu*** *(μ)* ***Cephei***, called the 'Garnet Star' by William Herschel with its striking red colour remains readily visible. The groups of stars, known as the ***Double Cluster*** in Perseus (NGC 869 & NGC 884, often known as h and χ Persei), lying between Perseus and Cassiopeia, are well-placed for observation.

Beyond the Milky Way, Perseus and Cassiopeia, the constellation of ***Andromeda*** is beginning to be lost in the north-western sky.

Castor and ***Pollux*** in ***Gemini*** are near the zenith for observers at 30°N, but for most northern observers the constellation is best seen when facing south. The head of ***Draco*** is now higher in the sky and easier to recognize, with its long, straggling 'body' curling round ***Ursa Minor*** and the Pole. ***Ursa Major*** has begun to swing round towards the north, and Cassiopeia is now lower in the western sky.

The ***Apollo 14*** lunar mission landed in the Fra Mauro formation on 5 February 1971.

For observers in the far north, most of ***Cygnus***, with its brightest star, ***Deneb*** (α Cygni), is visible in the north, and even ***Lyra***, with ***Vega*** (α Lyrae) may be seen at times. Most of the constellation of ***Hercules*** is visible, together with the distinctive circlet of ***Corona Borealis*** to its east. Observers farther south may see Deneb and even Vega peeping over the northern horizon at times during the night, although they will often be lost (like all the fainter stars) in the inevitable extinction along the horizon.

LOOKING SOUTH

Early February 23:00 — Mid February 22:00 — Late February 21:00

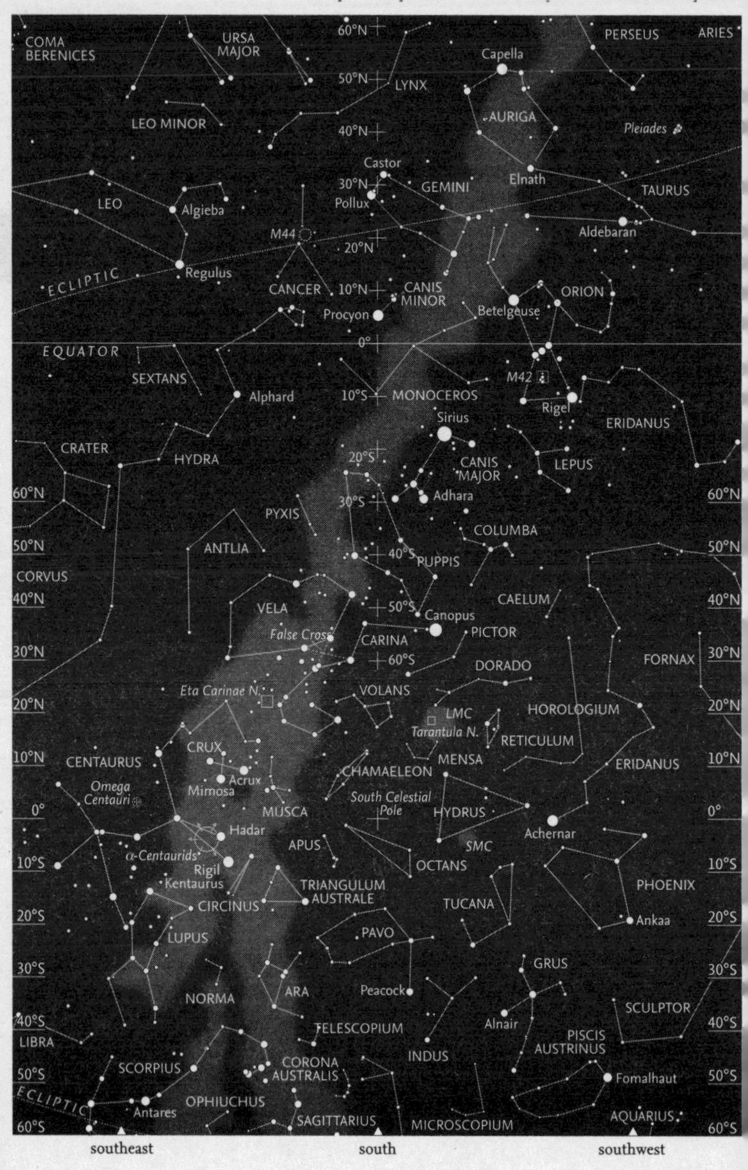

February – Looking South

Orion is now beginning to sink into the southwest, and the two brightest stars in the sky, ***Sirius*** and ***Canopus*** (α Carinae), are readily visible to observers at low northern latitudes and, of course, to those who are south of the equator. (Canopus is close to the zenith for those in the far south.)

South of ***Carina*** and the neighbouring constellation of ***Vela*** (both part of the original, and now obsolete, constellation of Argo Navis), lies the sprawling constellation of ***Centaurus***, surrounding the distinctive constellation of ***Crux***, the Southern Cross, which is on the horizon at 10°N. North of this, the 'False Cross', sometimes mistaken for the true constellation of Crux, consists of two stars from each of Vela (δ and κ Velorum) and Carina (ε and ι Carinae). The two brightest stars of Centaurus, ***Rigil Kentaurus*** (α Centauri) and ***Hadar*** (β Centauri) are slightly farther south, beyond Crux.

Sirius, α Canis Majoris (α Cma), in the southern celestial hemisphere, is the brightest star in the sky at magnitude -1.44.

The ***Large Magellanic Cloud*** (LMC) is almost on the meridian, early in the night, to the west of Carina and the small constellation of ***Volans***. It is surrounded by several small constellations: ***Mensa***, ***Reticulum*** and ***Dorado***. (Technically, it is largely within the area of Dorado.) Any optical aid (such as binoculars) will begin to show some of the remarkable structures within the LMC, including the great ***Tarantula Nebula***, or ***30 Doradus***, a site of active star formation.

Farther south and west lies ***Achernar*** (α Eridani) , the bright star at the end of ***Eridanus***, the long straggling constellation that represents a river and that may now be traced all the way from where it begins near ***Rigel*** (β Orionis) in Orion.

Beyond Crux, and on the other side of the Milky Way, lies the rest of Centaurus. Northeast of Crux is the finest and brightest globular cluster in the sky, ***Omega (ω) Centauri*** also known as NGC 5139. It is the largest globular cluster in our Galaxy and is estimated to contain about 10 million stars. Although appearing like a star, its non-stellar nature was discovered by Edmond Halley in 1677.

Carbonaceous chondrites

The meteorites classed as carbonaceous chondrites are the most valuable scientifically. They are considered to be extremely ancient – primitive – and are thought to closely resemble the material from which the Solar System, and all the bodies within it, origninally arose. They are considered to be residual portions of the original material, left over when the main portion accreted to form the planets. The carbonaceous chondrites form only a very small proportion (less than 5 per cent) of known meteorites, and only about 50 have actually been observed as falls. Observations of a fall – ideally with photographs of the trajectory in the atmosphere – are significant, because they may enable the object's orbit to be determined, and thus allow the original location of the meteorite within the Solar System to be established.

Many minor planets are known to consist of carbonaceous material and most of these are found in the outermost region of the Minor Planet Belt, which lies between the orbits of Mars and Jupiter. Based on the observed properties of various minor planets, it is possible to conclude that some classes of meteorites have originated in those particular minor planets. This is the case, for example, with the (non-corbonaceous) group known as howardite–eucrite–diogenites which were matched to the minor planet ***(4) Vesta.*** This relationship was confirmed by the ***Dawn*** spacecraft, which orbited the minor planet between 16 July 2011 and 5 September 2012. In the case of the carbonaceous chondrites, one group, known as CR meteorites, show a close match with the properties of minor planet ***(2) Pallas.***

The significance of carbonaceous chondritic meteorites lies in the fact that many contain amino acids, simple organic chemicals, and water. These compounds are essential for life, and there is a persistent opinion that life on Earth has arisen only because these materials have been delivered by carbonaceous meteorites or carbonaceous minor planets that have impacted on the Earth. Some groups of carbonaceous chondrites contain very high percentages of water – 'high' in this instance implying between 3 and 22 per cent. Many show evidence of being considerably altered by the presence of liquid water. The most famous carbonaceous chondrite meteorite is probably the Murchison meteorite, observed to fall near Murchison, Victoria, Australia on 28 September 1969. Both Murchison and the recent Winchcombe meteorite (page 67) belong to the group known as the CM meteorites. Their material appears to resemble that collected by the ***Hayabusa 2*** spaceprobe from the minor planet ***(162173) Ryugu***, and returned to Earth in December 2020. The Murchison meteorite has been particularly important and the subject of numerous, significant studies. It has been found to contain a phenomenal number of molecular compounds (at least 14,000), including some 70 amino acids. Some estimates put the number of potential compounds in the meteorite at hundreds of thousands, or even as high as one million. In January 2020, an international team of cosmochemists announced that some silicon carbide (SiC, carborundum) particles from the Murchison meteorite were the very oldest particles ever detected. They had anomalous isotopic ratios of silicon and carbon, implying that they were formed outside the Solar System. These grains have a suggested age of 7000 million years, some 2500 million years older than the Solar System itself.

A large fragment of the Murchison meteorite.

The ***Hayabusa 2*** spaceprobe collected its first sample from the minor planet ***(162173) Ryugu*** on 21 February 2019. A second sample was collected later.

March

March – Introduction

On Sunday, March 20, the Sun appears to cross the celestial equator from south to north. The point at which this occurs, known as the ***First Point of Aries***, was originally named many centuries ago when it lay in the constellation of ***Aries***, and now (because of the phenomenon of precession, which changes the direction of the Earth's axis of rotation) it actually lies in the constellation of ***Pisces***. It is close to the border with the constellation of ***Aquarius*** and lies southwest of the star λ Piscium. It is moving towards Aquarius at the rate of about one degree every 70 years, and will eventually enter that constellation in some hundreds of years. (*The Age of Aquarius*, made popular by the song, is an astrological idea, rather than a true astronomical concept and, astronomically, is rather premature.)

The planet Uranus was discovered on 13 March 1781, by ***William Herschel***, working from his house in Bath in Somerset.

Day and night at the equinoxes are almost exactly equal in length, and the northern hemisphere's season of spring is considered to begin in March. (Although theoretically the amount of time in day and night is the same, in fact refraction in the atmosphere 'raises' the position of the Sun at sunset and sunrise, so daylight lasts slightly longer than darkness.) The hours of daylight and night-time darkness change most rapidly around the equinoxes in March and September. The corresponding equinox in autumn (for the northern hemisphere, the spring equinox for those in the south), when the Sun crosses the celestial equator from north to south, occurs on September 23 in 2022, and always lies in the constellation of ***Virgo***.

The First Point of Aries is sometimes known as the Cusp of Aries and the other equinox, in September, as the Cusp of Virgo.

More information about the equinoxes is to be found on pages 80–81.

The planets

Mercury is too close to the Sun to be seen this month. ***Venus*** is visible in the morning sky, and is very bright (mag. -4.7 to -4.5). It is at western elongation on March 20 (the day of the equinox). ***Mars*** is also in the morning sky, initially in ***Sagittarius***, but moving east into ***Capricornus*** and fading slightly to mag. 1.1. ***Jupiter*** is at superior conjunction in ***Aquarius*** on March 5, but is too close to the Sun to be visible at any time during this month. ***Saturn*** is in ***Capricornus***, very low in the morning sky. ***Uranus*** remains in ***Aries*** at mag. 5.8. ***Neptune*** is at superior conjunction (and thus invisible) on March 13.

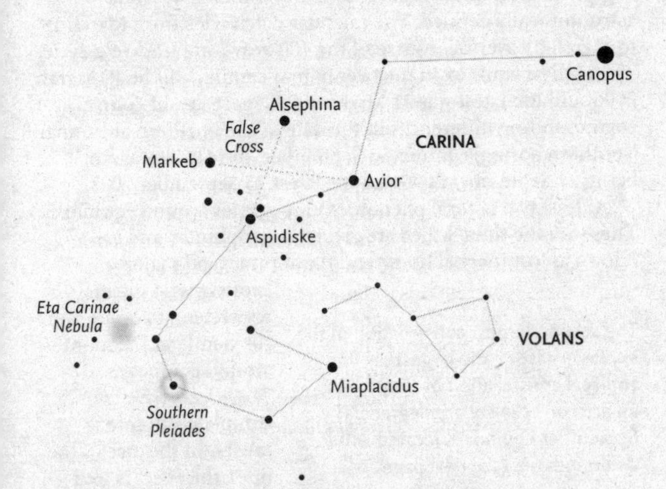

This month the constellation of Carina is well placed for observers at 20°N or more to the south. Canopus (α Car) is the second brightest star in the sky. Avior (ε Car) and Aspidiske (ι Car) form the False Cross, together with two stars that belong to the constellation of Vela: Alsephina (δ Vel) and Markeb (κ Vel). The small constellation of Volans is almost embedded in Carina.

The Equinoxes

Many early societies (most early civilizations were in the northern hemisphere) started their year, and the cycle of seasons, at the spring equinox. When Julius Caesar established what has come to be known as the Julian calendar (page 63) he set the equinox at March 25, and the year started on that date. Because a year in the Julian calendar is slightly longer than the year as determined by the Sun, the date of the equinox slowly became earlier and by 1580 had drifted back to March 21. Pope Gregory XIII instigated the reforms that led to what is now known as the Gregorian calendar, which he introduced in October 1582, and which we still use. Even so, there is a slight variation in the exact date of the equinox, which is astronomically defined. The calendar date varies from March 19 to March 21 over the course of the 400-year-long leap-year cycle. The earliest equinox in the twenty-first century will be 19 March 2096, and the latest was 21 March 2003. There are, of course, corresponding differences in the date of the northern autumnal (southern spring) equinox in September; the earliest, again, being 21 September 2096 and the latest 23 September 2003.

At least two natural phenomena are greater around equinoxes. These are the tides, which are greater in amplitude (and are known as 'equinoctial tides') and the aurorae (and major geomagnetic storms), which tend to occur more frequently in the months around equinoxes. The effects have two completely different causes. In the case of the tides, the effect is well-known and was actually explained by Isaac Newton in the late seventeenth century. Newton explained the effect by stating that the gravitational effect of the Sun is greatest when its declination (page 8) is at a minimum. It is then acting along a line to the centre of the Earth, and is not offset to north or south. The Sun's declination is precisely zero when it crosses the celestial equator, and this occurs, twice a year, at the equinoxes. A somewhat similar effect

The second largest constellation in the sky (after Hydra, the largest), is the zodiacal constellation of ***Virgo***, with an area of 1294 square degrees. The September equinox is located within its boundaries (see next page).

occurs with the Moon, but because of the inclination of the Moon's orbit and the rotation of the Earth, the results of the changes in the Moon's declination are extremely complex. Even the difference in the Moon's declination over the course of a day has an effect. This daily change causes a difference in the height of the tides. In general, these alternate in height with a period of 12 hours, giving rise to what is known as the semi-diurnal tide. More complex effects occur through the changes in the orientation of the Moon's orbit around the Earth, which occur with a very long period, measured in years.

The causes of the increase in frequency of aurorae and geomagnetic storms in the months around the equinoxes are still poorly understood. The effect is definite however, and is known as the Russell-McPherron effect. The Sun's axis of rotation is inclined at 7°.25 to the ecliptic (the plane of the Earth's orbit around the Sun). At certain times of the year – which happen to coincide with the equinoxes – the north or south poles are tilted towards the Earth. The south pole of the Sun is tilted towards the Earth in March, and its north pole in September. The magnetic field lines around the poles are open to space – they extend out into the interplanetary magnetic field. At other latitudes (closer to the Sun's equator) the magnetic field lines loop back into the surface. Around the equinoxes there is thus a more direct connection between the magnetic fields at the poles of the Sun and Earth. It is therefore easier for any material ejected from the Sun to travel along the magnetic field lines, reach the Earth and affect its magnetic field, either in the form of a major geomagnetic storm or as aurorae.

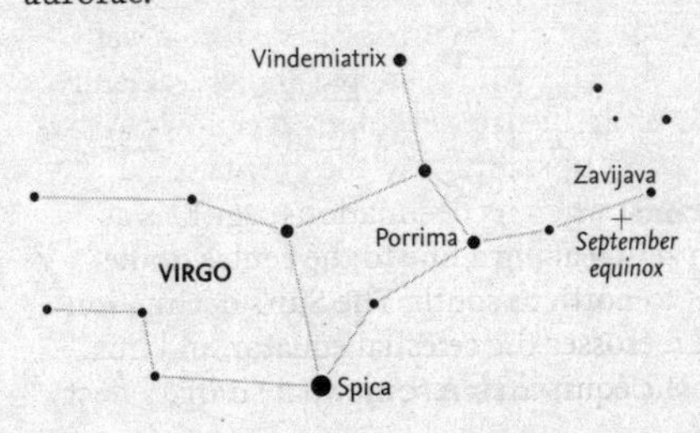

The constellation of Virgo, showing the location of the September equinox.

Sunrise and sunset

City	Date	Sunrise	Sunset
Buenos Aires, Argentina			
	Mar. 01	09:41	22:30
	Mar. 31	10:05	21:49
Cape Town, South Africa			
	Mar. 01	04:34	17:22
	Mar. 31	04:58	16:43
London, UK			
	Mar. 01	06:46	17:41
	Mar. 31	05:39	18:33
Los Angeles, USA			
	Mar. 01	14:22	01:49
	Mar. 31	13:42	02:13
Nairobi, Kenya			
	Mar. 01	03:41	15:49
	Mar. 31	03:34	15:40
Sydney, Australia			
	Mar. 01	19:43	08:32
	Mar. 31	20:07	07:52
Tokyo, Japan			
	Mar. 01	21:10	08:36
	Mar. 31	20:28	09:01
Washington, DC, USA			
	Mar. 01	11:41	23:01
	Mar. 31	10:54	23:31
Wellington, New Zealand			
	Mar. 01	18:03	07:05
	Mar. 31	18:36	06:15

NB: the times given are in Universal Time (UT)

The Moon's phases and ages

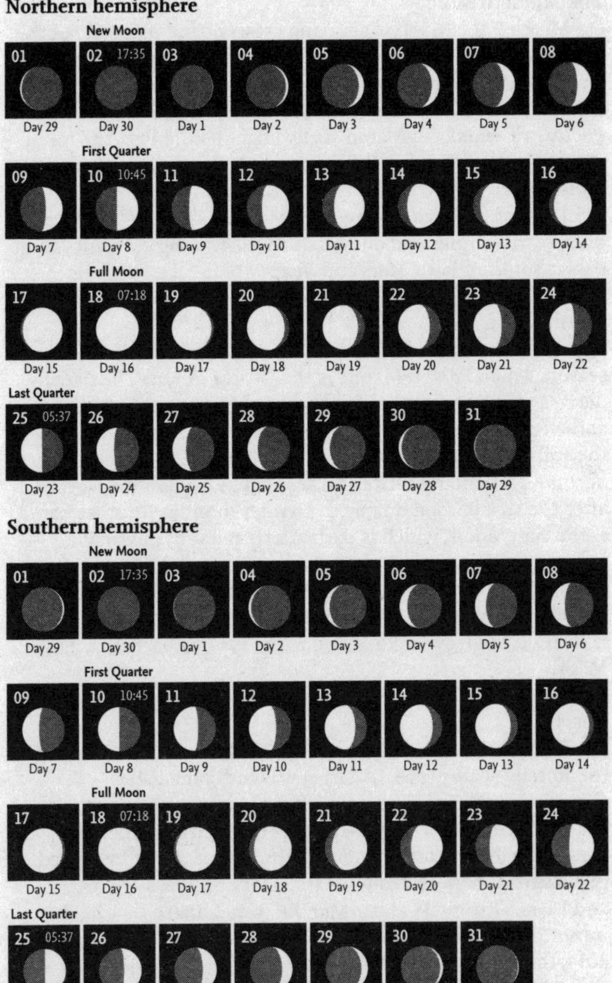

The Moon

The Moon in March

On March 7 the Moon is a waxing crescent and just 0.8° south of ***Uranus*** (mag. 5.8) in ***Aries.*** On March 9, just before First Quarter, it is 7.0° north of orange-tinted ***Aldebaran*** in ***Taurus.*** On March 13, the waxing gibbous Moon passes 2.4° south of ***Pollux*** in ***Gemini.*** On March 16, two days before Full, the Moon is 4.9° north of ***Regulus.*** It then passes between that star and ***Algieba*** (γ Leonis). On March 20, two days after Full Moon, it is 5.1° north of ***Spica*** in ***Virgo.*** On March 23 it is 3.2° north of ***Antares*** in ***Scorpius*** and on March 28 the waning crescent Moon passes south of both ***Mars*** and ***Venus.***

Worm Moon

One common name for the last Full Moon of the winter season, which falls in March, is the 'Worm Moon'. The name derives from the fact that earthworms become active in the soil at the end of winter and are sometimes seen at the surface of the soil. Other names are the 'Crow Moon', because the birds become particularly active and are avid to feed on the worms, after the lack of food during the winter months. Another name is the 'Sap Moon', which is particularly relevant in Canada, because this is the time when maple trees may be tapped for the sap (to produce the maple syrup beloved by Canadians). In Europe, the term 'Lenten Moon' was sometimes used, and this is the Old English/Anglo-Saxon name for this particular Full Moon.

Rosetta

On 2 March 2004, the European Space Agency (ESA) launched the ***Rosetta*** spaceprobe. Its target was the comet 67P/Churyumov–Gerasimenko, which it reached on 6 August 2014, entering orbit on 10 September 2014. The comet passed perihelion on 12 August 2015. The spacecraft received gravitational corrections to its trajectory by flybys of Earth and Mars. During its journey to the comet it also flew by the minor planets (21) Lutetia and (2867) Šteins. On 12 November 2014, the lander module ***Philae*** became the first object ever to land on the nucleus of a comet. Unfortunately, because of

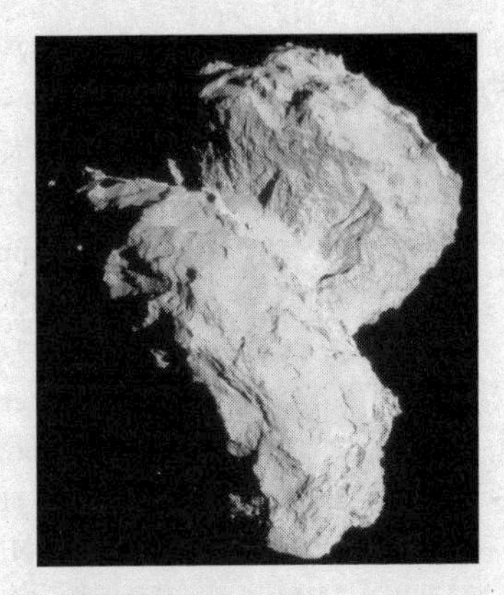

An image of 67P/Churyumov–Gerasimenko obtained by Rosetta when the spaceprobe was about 64 km from the comet.

the comet's low gravity, Philae bounced across the surface and landed, wedged in a position where it was unable to carry out many of its functions, although some useful data was later derived from it. *Rosetta* remained with the comet until the end of the mission, when it was deliberately landed on the surface on 30 September 2016.

One of the most significant findings from *Rosetta*'s scientific work was the discovery that in the water vapour in the comet's coma the ratio of deuterium (heavy hydrogen, where there is an additional proton in the nucleus of the hydrogen atoms, giving two protons) to ordinary (light) hydrogen (with a single proton) was as much as three times as high as the ratio found in terrestrial water. This suggests that water on Earth was not derived from comets such as Churyumov–Gerasimenko, as has been commonly believed.

Another finding was that this particular comet contained many solid particles; so many in fact that rather than being a 'dirty snowball' it could be likened to an 'icy dustball'. There were also strong indications that the body consisted of numerous individual portions that had accumulated together, with there being considerable porosity within the comet, rather than it being a single, consolidated body.

The comet consists of two distinct lobes, and this suggests that it is what is known as a 'contact binary'; that it has arisen as the result of a low-velocity collision between two separate bodies.

Comet 67P/Churymov-Gerasimenko reaches perihelion on 2 November 2021.

M

Calendar for March

02	13:00 *	Mercury 0.7°S of Saturn
02	17:35	New Moon
02	18:35	Jupiter 4.1°N of Moon
03	09:02	Neptune 3.7°N of Moon
05	14:07	Jupiter at superior conjunction
07	06:08	Uranus 0.8°N of Moon
09	16:59	Aldebaran 7.0°S of Moon
10	10:45	First Quarter
10	23:04	Moon at apogee = 404,268 km
12	14:00 *	Mars 4.0°S of Venus
13	07:34	Pollux 2.4°N of Moon
13	11:43	Neptune at superior conjunction
14–15		γ-Normid meteor shower maximum
16	02:00	Regulus 4.9°S of Moon
18	07:18	Full Moon
20	02:24	Spica 5.1°S of Moon
20	09:25	Venus at greatest elongation (46.6°W, mag. -4.5)
20	15:33	March equinox
23	11:16	Antares 3.2°S of Moon
23	23:37	Moon at perigee = 369,760 km
25	05:37	Last Quarter
28	02:53	Mars 4.1°N of Moon
28	09:49	Venus 6.7°N of Moon
28	11:41	Saturn 4.4°N of Moon
29	13:00 *	Saturn 2.2°S of Venus
30	14:36	Jupiter 3.9°N of Moon
30	19:18	Neptune 3.7°N of Moon

* *These objects are close together for an extended period around this time.*

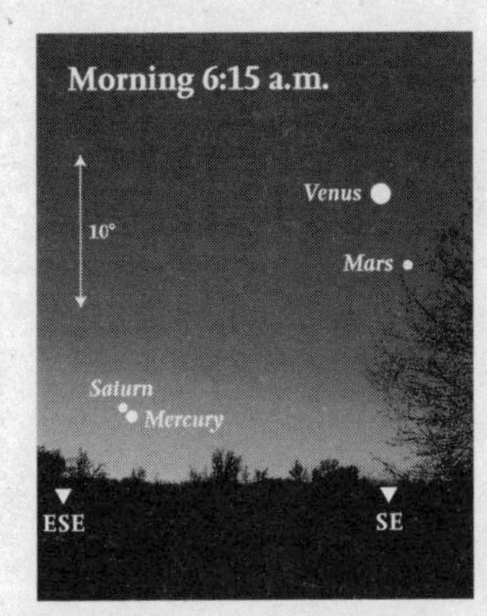

March 2 • *Venus and Mars are in the southeast. Saturn and Mercury are closer to the horizon (as seen from central USA).*

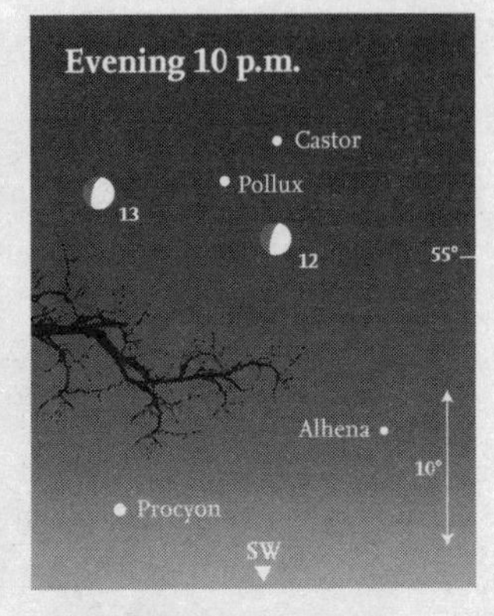

March 12–13 • *The waxing gibbous Moon passes below Castor and Pollux (as seen from London).*

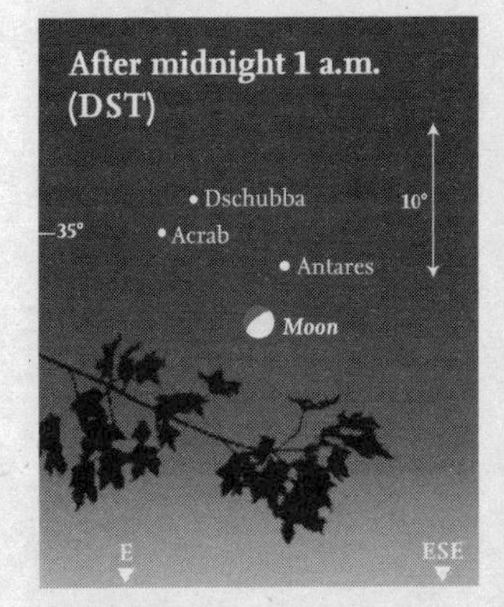

March 24 • *The Moon is close to Antares with Acrab (β Sco) and Dschubba (δ Sco) nearby (as seen from Sydney).*

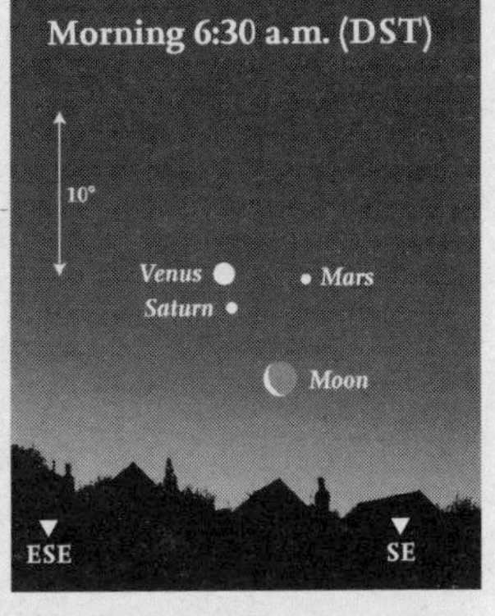

March 28 • *The crescent Moon in the company of three planets: Mars, Saturn and bright Venus (as seen from central USA).*

M

Early March 23:00 — Mid March 22:00 — Late March 21:00

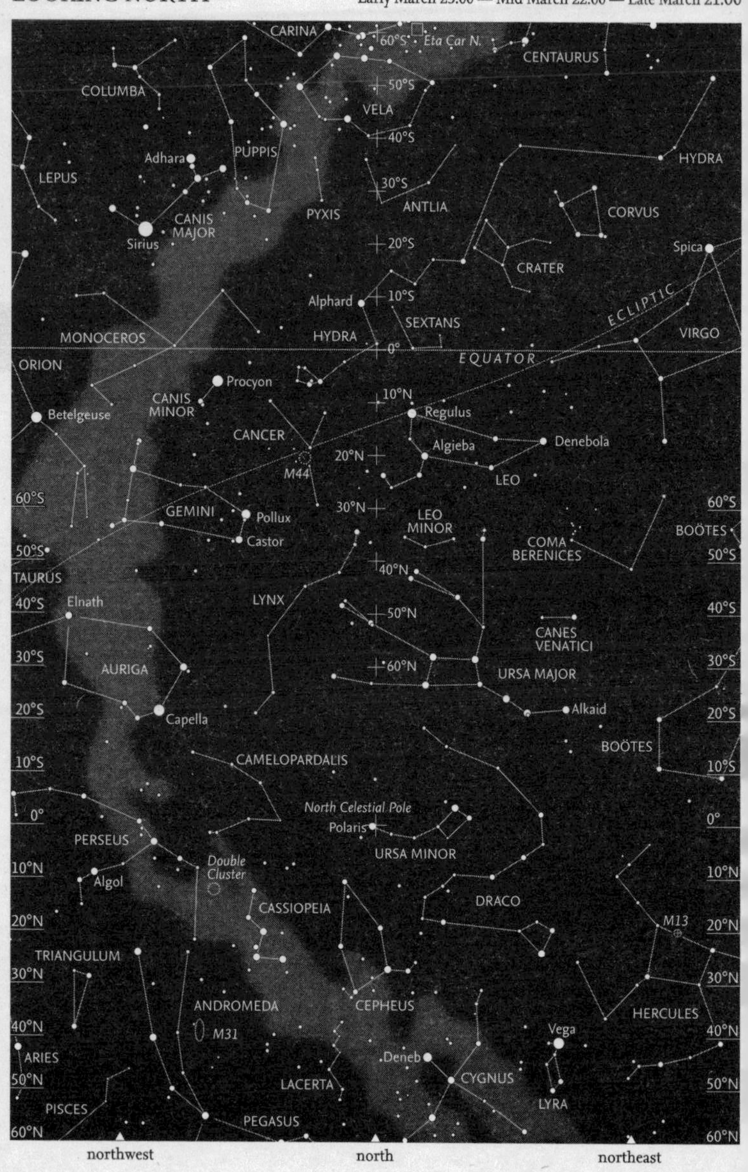

March – Looking North

The highly distinctive (and widely recognized) constellation of ***Ursa Major*** with the distinctive asterism of the Plough (or Big Dipper) is now 'upside down' and near the zenith for observers in the far north, for whom it is particularly difficult to observe. At this time of year, it is high in the sky for anyone north of the equator. Only observers farther towards the south will find it lower down towards their northern horizon and reasonably easy to see. However, at 30°S, even the seven stars making up the main, easily recognized portion of the constellation are too low to be visible.

M

Auriga, with brilliant ***Capella*** (α Aurigae) is also very high on the opposite side of the meridian. The constellation of ***Perseus*** lies between it and ***Andromeda*** on the western side of the sky.

Ursa Minor, also with seven main stars, one of which is ***Polaris***, the Pole Star, and the long constellation of ***Draco*** that winds around the Pole, are readily visible for anyone in the northern hemisphere, although, of course, Polaris is right on the horizon for anyone at the equator, and thus always lost to sight. ***Cepheus*** is near the meridian to the north, with ***Cassiopeia***, to its west beginning to turn and resume its 'W' shape. The constellation of Andromeda is now diving down into the northwestern sky. In the east, beyond ***Alkaid*** (η Ursae Majoris), the final star in the 'tail' of Ursa Major, lies the top of ***Boötes***. Farther to the south, most of ***Hercules*** and the 'Keystone' shape that forms the major portion of the body is visible.

The ***Pioneer 10*** spaceprobe was launched on 2 March 1972. It became the first probe to cross the asteroid belt and the first to photograph Jupiter, beginning on 6 November 1973.

Observers at 50°N may occasionally be able to detect bright ***Deneb*** (α Cygni) and ***Vega*** (α Lyrae) skimming the horizon, together with portions of those particular constellations, although most of the time they will be lost in the extinction that occurs at such low altitudes.

Early March 23:00 — Mid March 22:00 — Late March 21:00

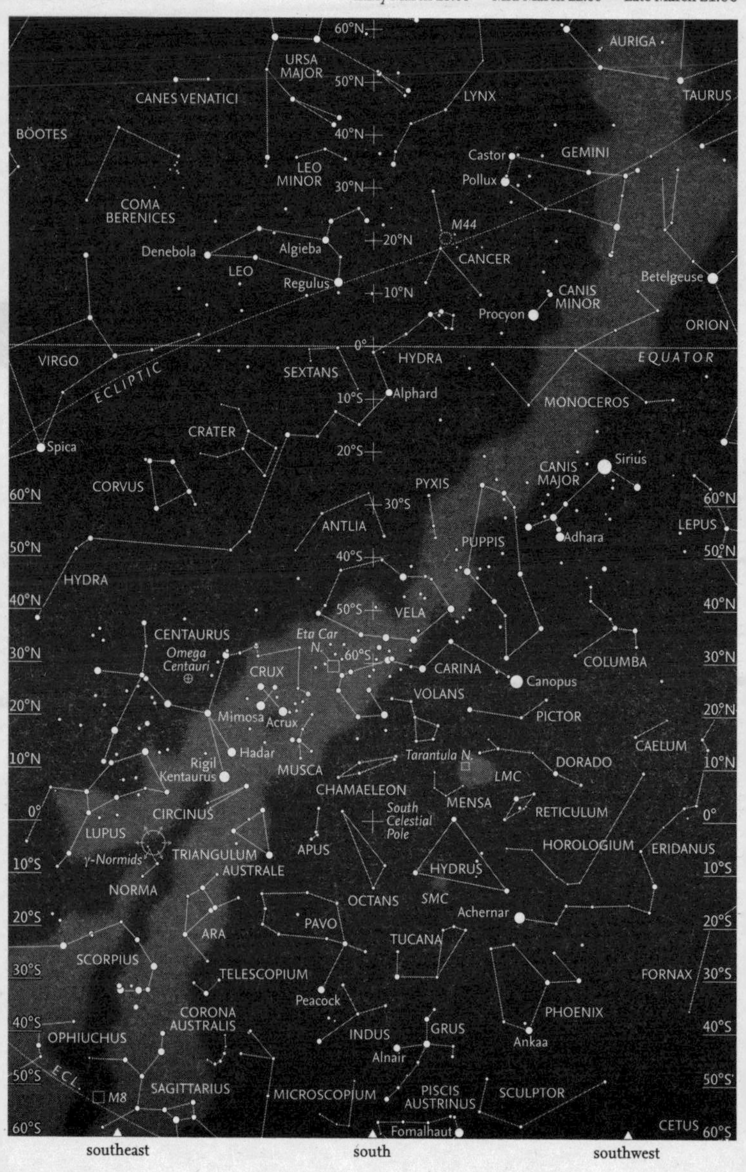

March – Looking South

The distinctive constellation of ***Leo*** is close to the meridian early in the night, clearly visible for anyone north of the equator. It is easily recognized, with bright ***Regulus*** (α Leonis) at the base of the 'backward question mark' of the asterism known as 'the Sickle', which lies north of Regulus. To the west of Leo is the inconspicuous constellation of ***Cancer***, and still farther away from the meridian, the far more striking constellation of ***Gemini***, with the bright stars ***Castor*** (α Geminorum), the closer to the Pole, and ***Pollux*** (β Geminorum). The constellation straddles the ecliptic, and Pollux may sometimes be occulted by the Moon (as may Regulus), although no such occultation occurs in 2022. Castor is remarkable in that even a fairly small telescope will show it as consisting of three stars (two fairly bright, and one fainter). However, more detailed investigation reveals that each of those stars is actually a double, so the whole system consists of no fewer than six stars.

Below Cancer is the very distinctive asterism of the 'Head of Hydra', consisting of five (or six) stars, that is the western end of the long constellation of ***Hydra***, the largest constellation in the sky, that runs far towards the east, roughly parallel to the ecliptic. ***Alphard*** (α Hydrae) is south, and slightly to the west of Regulus in Leo and is relatively easy to recognize as it is the only fairly bright star in that region of the sky. North of Hydra and between it and the ecliptic and the constellation of ***Virgo*** are the two constellations of ***Crater*** and ***Corvus***. Farther west, the small constellation of ***Sextans*** lies between Hydra and Leo.

Farther south, the Milky Way runs diagonally across the sky, and the constellation of ***Vela*** straddles the meridian. Slightly farther south is the constellation of ***Carina***, with, to the west, brilliant ***Canopus*** (α Carinae), which lies below the constellation of ***Puppis***, which is itself between Vela and ***Canis Major*** in the west.

Crux (the Southern Cross) is southeast of Carina and the two principal stars of ***Centaurus***, ***Rigil Kentaurus*** (α Centauri) and ***Hadar*** (β Centauri). The ***Large Magellanic Cloud*** (LMC) lies west of these stars, on the other side of the meridian.

Easter Sunday

Easter Sunday is calculated to occur after the Full Moon following March 21. Now that astronomers know the phases of the Moon decades or hundreds of years in advance, you might think that knowing the date of Easter would be easy. But it is not as easy as that. The Christian religious festival is calculated using a lunar calendar: the eclesiastical calendar. In this, the 'months' alternate between 30 and 29 days. To astronomers, the time between New Moon and New Moon (known as the synodic month) averages 29.530 587 981 days (to take it to nine decimal places). That is just slightly more than twenty-nine and a half days, so 'months' that alternate between 29 and 30 days do give a reasonable agreement. (But not always, see the description of the term 'Black Moon' on page 66.) In the ecclesiastical calendar, Full Moon is always taken to be on the 14th day of the lunar month, reckoned in local time. So the ecclesiastical calendar may get out of step with the astronomical calendar, in which Full Moon is defined as occurring at a specific date and time (in Co-ordinated Universal Time, UTC). What is more, to astronomers, the exact date and time of Full Moon apply worldwide. In 2022, the religious Easter Sunday is on April 17 (one day after Full Moon), whereas if it were calculated astronomically it would (for once) be the same day.

April

April – Introduction

There are no major astronomical events in April 2022, but three meteor showers are active, beginning mid-month. Two are best seen from the northern hemisphere, but there is one significant southern shower. The one moderate, northern shower is the ***Lyrids*** (often called the ***April Lyrids*** to distinguish them from several minor showers that originate in the constellation at various times during the year). Regrettably, in 2022 the shower begins on April 14, two days before Full Moon, and comes to a weak maximum of some 18 meteors per hour on April 22–23, at Last Quarter, so conditions in 2022 are generally unfavourable. There is another, stronger shower, with a possible hourly rate of 50 meteors per hour. This is the ***η-Aquariids***. This shower begins on April 19, three days after Full Moon when it is waning gibbous, and continues over New Moon on April 30, to peak on May 6, continuing well into May (May 28). Conditions for observing this shower

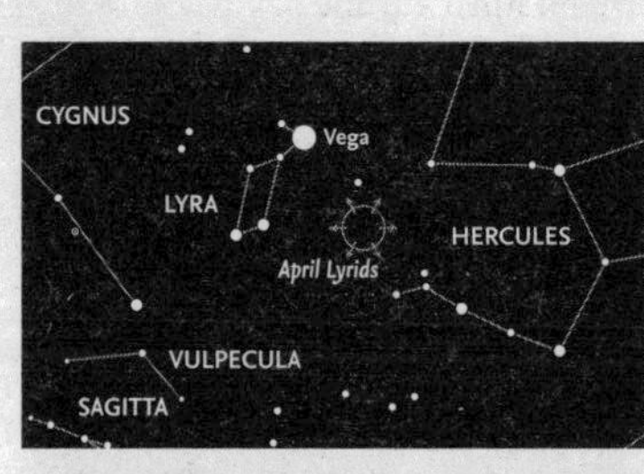

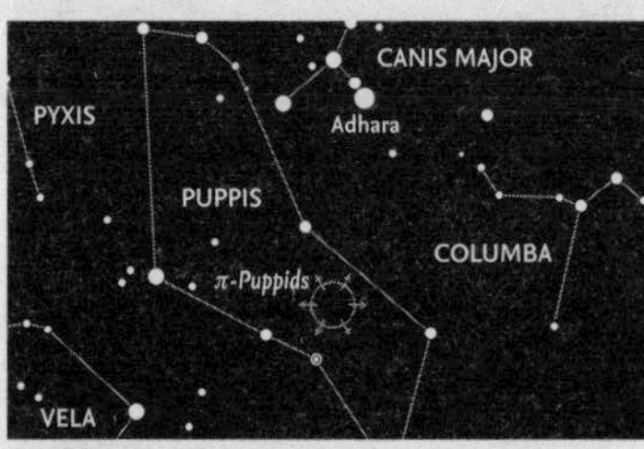

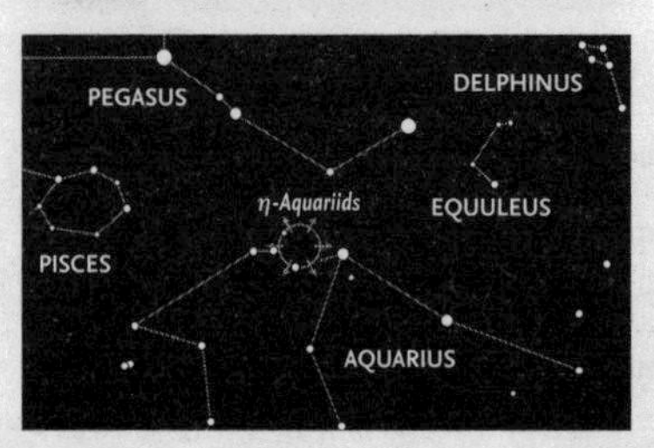

The charts show the location of the April Lyrids radiant (top), the π-Puppids radiant (middle) and the η-Aquariids radiant (bottom).

are poor at the beginning, improve around maximum, but then deteriorate again around Full Moon on May 16. The η-Aquariid shower, like the Orionids of October, is the result of debris left in orbit behind the famous comet 1P/Halley. This is a moderately predictable shower, and the maximum rate is between 40 and 50 meteors per hour. The radiant for this shower is close to the 'Y'-shaped asterism known as the 'Water Jar' in ***Aquarius***.

The southern meteor shower, the ***π-Puppids***, begins in April. This shower starts to be active on April 15, one day before Full Moon, and lasts until April 28, with maximum on the night of April 23–24. The shower was unknown until 1972. The rate seems to be rather variable, reaching a maximum of about 40 meteors per hour in 1977 and 1983, and it is difficult to predict whether many meteors will be seen. As with the Lyrids, maximum occurs when the Moon is around Last Quarter. The parent comet is believed to be Comet 26P/Grigg–Skjellerup.

The nearly disastrous **Apollo 13** manned lunar mission was launched on 11 April 1970. Two days later a ruptured oxygen tank caused the mission to be aborted. After looping round the Moon, and enduring extremely dangerous conditions, the three astronauts landed safely on 17 April 1970.

The planets

Most of the planets are low in the sky this month. ***Mercury*** is largely invisible. It is at superior conjunction on the far side of the Sun on April 2, but comes to greatest eastern elongation (20.9° from the Sun at mag. 0.2) in the evening sky on April 29. ***Venus*** is bright (mag. -4.5 to -4.1) but very low in the morning sky. ***Mars*** (mag. 1.1 to 0.9) moves from ***Capricornus*** into ***Aquarius***, but is also low in the morning sky. On April 4, Mars is 0.3° south of ***Saturn*** and both planets are of similar magnitude (mag. 1.1 and 0.9, respectively). ***Jupiter***, again, is very low in the morning sky (fading slightly from mag. -2.1 over the month). It is initially just within ***Aquarius***, and then moves into ***Pisces***. ***Saturn*** (mag. 0.9) is in Capricornus. ***Uranus*** (mag. 5.9) is slowly moving eastwards in ***Aries***. ***Neptune***, in Pisces, is mag. 8.0 to 7.9.

Sunrise and sunset

City	Date	Sunrise	Sunset
Buenos Aires, Argentina			
	Apr.01	10:06	21:48
	Apr.30	10:29	21:12
Cape Town, South Africa			
	Apr. 01	04:58	16:41
	Apr. 30	05:20	16:06
London, UK			
	Apr. 01	05:37	18:34
	Apr. 30	04:35	19:23
Los Angeles, USA			
	Apr. 01	13:40	02:13
	Apr. 30	13:05	02:36
Nairobi, Kenya			
	Apr. 01	03:34	15:39
	Apr. 30	03:28	15:32
Sydney, Australia			
	Apr. 01	20:08	07:51
	Apr. 30	20:29	07:16
Tokyo, Japan			
	Apr. 01	20:27	09:02
	Apr. 30	19:50	09:26
Washington, DC, USA			
	Apr. 01	10:53	23:32
	Apr. 30	10:12	24:00
Wellington, New Zealand			
	Apr. 01	18:37	06:13
	Apr. 30	19:08	05:29

NB: the times given are in Universal Time (UT)

The Moon's phases and ages

Northern hemisphere

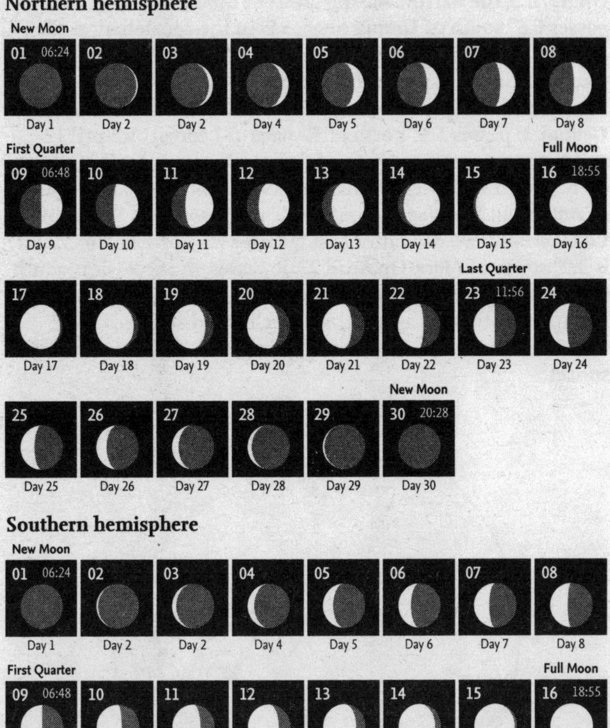

Southern hemisphere

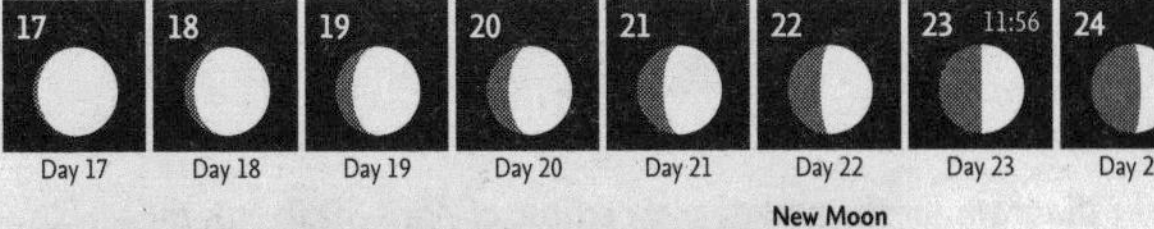

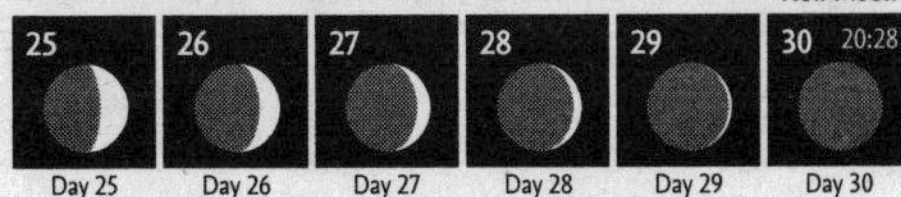

A

The Moon

On April 3, the narrow waxing crescent Moon (two days old) passes 0.6° south of ***Uranus*** (mag. 5.8) in the constellation of ***Aries*** in the evening sky. On April 6, when the Moon is still a waxing crescent, it passes 7.2° north of orange ***Aldebaran*** in ***Taurus.*** At First Quarter on April 9, it is 2.2° south of ***Pollux*** in ***Gemini.*** It passes 5.1° north of ***Regulus*** (α Leonis) on April 12 and the same distance north of ***Spica*** in ***Virgo*** on April 16. (Both events occur in daylight.) On April 19, the Moon is 3.1° north of ***Antares*** in ***Scorpius*** and on April 24, 4.5° south of ***Saturn*** in ***Capricornus.*** Two days after Last Quarter, on April 25, the Moon is 3.9° south of ***Mars.*** On April 27, it passes, in succession, south of ***Venus***, ***Neptune*** and ***Jupiter***, very low in the morning sky. On April 30, there is a partial solar eclipse, visible only from Chile and Argentina. Greatest eclipse is at 20:41 UT, just before sunset.

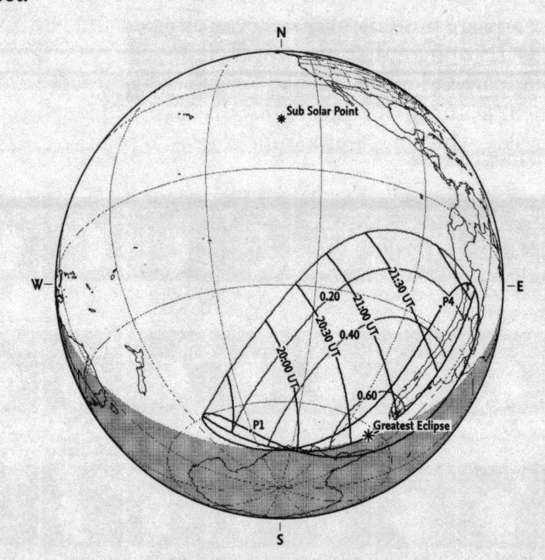

As the diagram for the partial solar eclipse of April 30 shows, the greatest eclipse actually occurs out at sea, south of Cape Horn at the very tip of South America, and just offshore from the Antarctic Peninsula. The maximum magnitude of the eclipse is about 64 per cent.

Pink Moon

The Full Moon in April is often known as the Pink Moon. The term derives from the pink flowers, particularly the plants known as phlox, that are prominent in early spring. Other names for this Full Moon are 'Sprouting Green Moon' (from the vibrant green shoots that appear in spring), 'Fish Moon' (because it becomes possible to catch fish again, once the rivers were no longer frozen), and 'Hare Moon' (hares are active during this month). Old-World terms were 'Egg Moon' – eggs were (and are) associated with Easter, and in particular, 'Pascal Moon' (because it was used to calculate the date of Easter – see page 92).

The lunar far side

The far side of the Moon is very different from the near side (the side always turned to Earth). Unlike the near side, where maria are prominent, there is an almost complete lack of maria or mare-like areas, or even flooded craters. The centre of Mare Orientale and Mare Moscoviense are the only prominent features, with one highly distinctive, flooded crater, Tsiolkovskiy, which has a pronounced central peak.

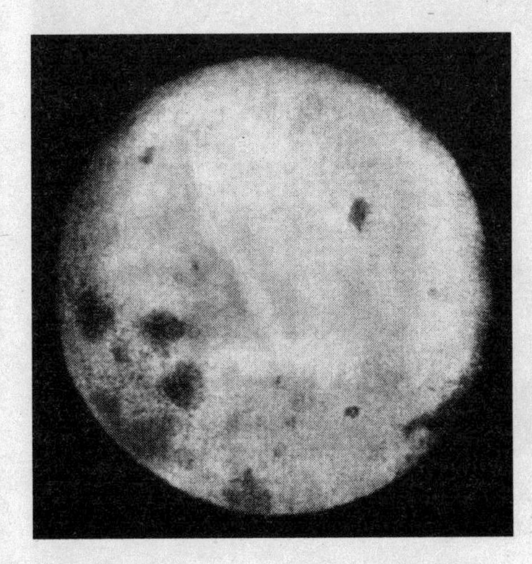

The first image of the lunar far side, as returned by the Luna 3 spaceprobe. Part of the near side is visible on the left.

The Moon appears to 'rock' backwards and forwards (and up and down), in the effect known as libration. This is caused by the result of the inclination of the Moon's orbit and its eccentricity (its deviation from circularity). About 18 per cent of the far side is visible on occasions, but some 82 per cent is permanently invisible from Earth. The first photographs of the far side were obtained in 1959 by the Soviet probe, *Luna 3*. These initial photographs were indistinct, by modern standards, but they indicated that the far side was very different from the near side.

A modern image of the lunar far side, created from material obtained by the Lunar Reconnaissance Orbiter spaceprobe.

That first image showed just a few mare areas that appear on the Moon's limb as seen from Earth (visible on the left-hand side). The rest of the image is essentially free from mare regions. The only distinct dark areas are those later named Mare Moscoviense (towards top right) and the crater Tsiolkovskiy, with its floor flooded by lava and with a prominent central peak (towards bottom right). Otherwise, the surface is covered in craters, like the highlands on the near side of the Moon. The difference is partly caused by the fact that the crust on the far side is much thicker than on the near side. Impacts did not penetrate deep enough to allow molten magma, which was located at depth, to rise to the surface. The Moon's centre of mass is not truly central, but is displaced towards the near side, producing a thin crust on the near side and a thick one on the far side. This asymmetry in the location of the Moon's centre of mass is the main reason for the tidal locking and synchronous rotation that always turns the same face towards Earth. The thin crust on the near side allowed impacts to penetrate deeply enough to allow the molten magma to rise to the surface, spread out as floods of lava and create the maria.

On 6 April 1973, the ***Pioneer 11*** spaceprobe was launched. It became the first probe to study Saturn.

On 20 April 1967, the ***Surveyor 3*** spaceprobe (later visited by Apollo 12 astronauts) landed in Oceanus Procellarum.

The ***Hayabusa 2*** spaceprobe fired a projectile into minor planet ***(162173) Ryugu*** on 5 April 2019 to create a crater. It collected a sample of the underlying material on 11 July 2019.

A

Calendar for April

01	00:25	Mercury 2.6°N of Moon
01	06:24	New Moon
02	23:11	Mercury at superior conjunction
03	17:27	Uranus 0.6°N of Moon
04	22:00 *	Mars 0.3°S of Saturn
06	01:16	Aldebaran 7.2°S of Moon
07	19:11	Moon at apogee = 404,438 km
09	06:48	First Quarter
09	15:50	Pollux 2.2°N of Moon
12	11:01	Regulus 5.1°S of Moon
12	20:00 *	Neptune 0.1°S of Jupiter
14–30		April Lyrid meteor shower
15–28		π-Puppid meteor shower
16	11:19	Spica 5.1°S of Moon
16	18:55	Full Moon
18	14:00 *	Uranus 2.1°S of Mercury
19–May.28		η-Aquariid meteor shower
19	15:13	Moon at perigee = 365,143 km
19	18:08	Antares 3.1°S of Moon
22–23		April Lyrid meteor shower maximum
23	11:56	Last Quarter
23–24		π-Puppid meteor shower Maximum
24	20:55	Saturn 4.5°N of Moon
25	22:05	Mars 3.9°N of Moon
27	01:51	Venus 3.8°N of Moon
27	03:20	Neptune 3.7°N of Moon
27	08:26	Jupiter 3.7°N of Moon
27	19:00 *	Neptune 0.01°N of Venus
29	08:09	Mercury at greatest elongation (20.6°E, mag. 0.2)
30	19:00 *	Venus 0.3°S of Jupiter
30	20:28	New Moon
30	20:41	Partial solar eclipse

** These objects are close together for an extended period around this time.*

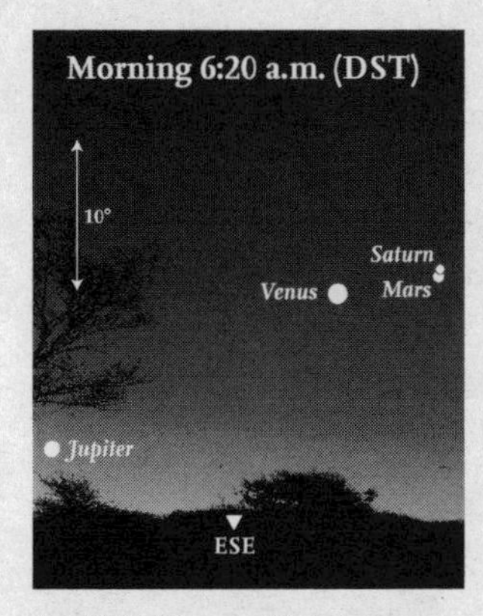

April 5 • *Venus with Saturn and Mars to the right (south). Jupiter is close to the horizon and farther east (as seen from central USA).*

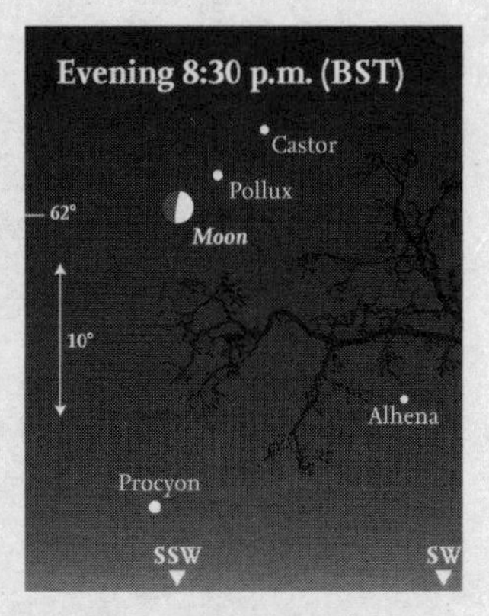

April 9 • *The First Quarter Moon is lining up with Castor and Pollux. Procyon and Alhena are nearby (as seen from London).*

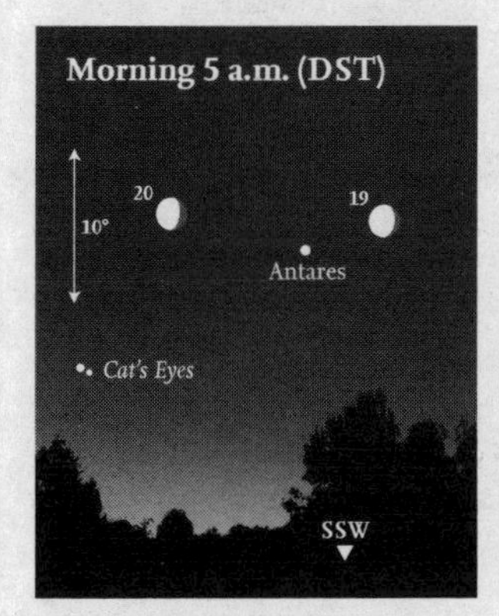

April 19–20 • *The Moon passes Antares, with the Cat's Eyes (λ and υ Sco) farther south (as seen from central USA).*

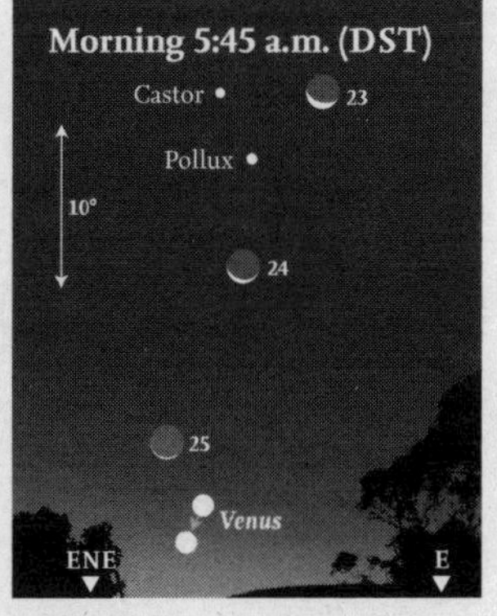

April 23–25 • *The crescent Moon passes Castor and Pollux. On April 25 it is close to Venus (as seen from central USA).*

LOOKING NORTH

Early April 23:00 — Mid April 22:00 — Late April 21:0

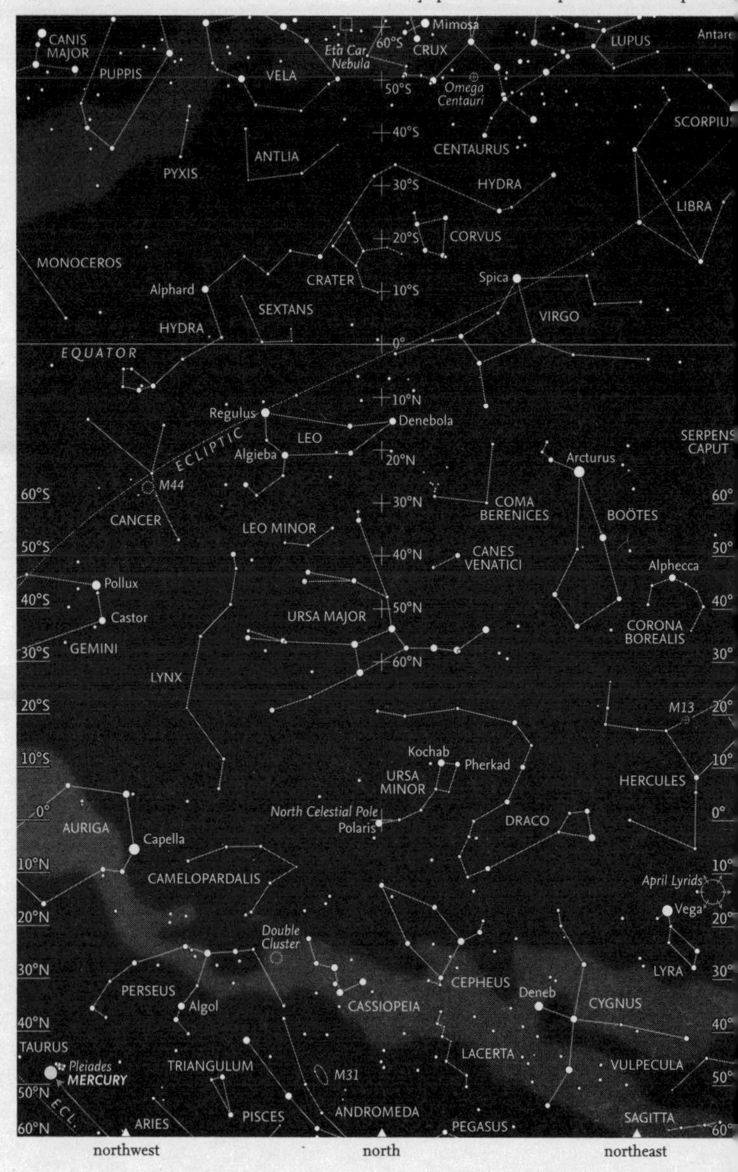

April – Looking North

Ursa Major is still high overhead for most northern-hemisphere observers, with ***Boötes*** and bright ***Arcturus*** (α Boötis), the star indicated by the 'arc' of the 'tail' of Ursa Major, to its northeast. The small circlet of ***Corona Borealis***, with its single bright star of ***Alphecca*** (α CrB) is still farther to the east. ***Cassiopeia*** has now swung round, and is almost on the meridian to the north, below ***Polaris*** and northwest of ***Ursa Minor***. ***Cepheus*** is beginning to climb higher and the head of ***Draco*** is almost due east of Polaris, but slightly farther south than the two 'Guards' in Ursa Minor, ***Kochab*** (β UMi) and ***Pherkad*** (γ UMi). On the other side of the meridian to the northwest lies the constellation of ***Auriga***, with bright ***Capella***. The inconspicuous constellation of ***Camelopardalis*** lies between Polaris, Auriga and ***Perseus***. The stars of the Milky Way run through Auriga, Perseus, and Cassiopeia towards the much more densely populated regions in ***Cygnus*** and farther south. Nearly the whole of Cygnus is visible above the horizon in the northeast, where the small constellation of Lyra, with the bright star ***Vega***, is clearly seen with ***Hercules*** above it.

The constellation of Perseus is beginning to descend into the northwest, following the constellation of ***Andromeda***, which, for most observers, has now disappeared below the northwestern horizon.

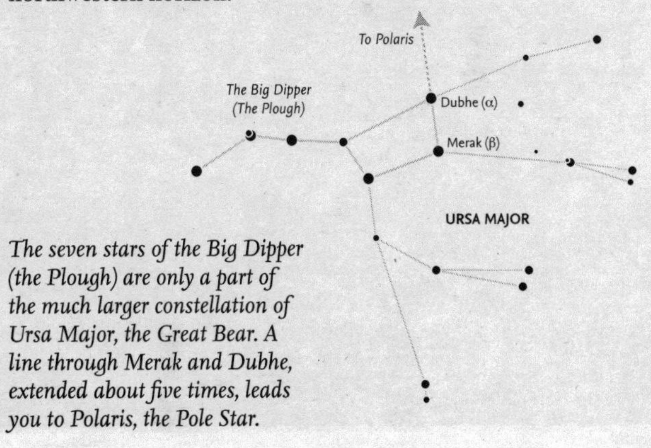

The seven stars of the Big Dipper (the Plough) are only a part of the much larger constellation of Ursa Major, the Great Bear. A line through Merak and Dubhe, extended about five times, leads you to Polaris, the Pole Star.

Early April 23:00 — Mid April 22:00 — Late April 21:0

April – Looking South

The constellation of ***Leo*** is still conspicuous in the south, although now it is ***Denebola*** (β Leonis) rather then ***Regulus*** that is close to the meridian. ***Boötes*** and ***Arcturus*** (α Boötis), the brightest star in the northern hemisphere, are high in the southeast. Beyond Leo, along the ecliptic to the east, the zodiacal constellation of ***Virgo*** is clearly visible, although ***Spica*** (α Virginis) is the only really bright star in the constellation.

Farther south, ***Crux*** is now nearly 'upright' and close to the meridian, with ***Rigil Kentaurus*** and ***Hadar*** (α and β Centauri, respectively), conspicuous to its southeast. Rigil Kentaurus is a multiple system, with two stars fairly easily visible with a small telescope. A third star in the system, known as ***Proxima Centauri***, is much fainter, and is actually the closest star to the Solar System at a distance of about 4.3 light-years. In recent years, it has been found to host the closest known exoplanet, which orbits Proxima every 11.2 days.

The stars of ***Vela*** and ***Carina*** lie to the west of Crux and the meridian, while much farther south, and out to the west is ***Canopus*** (α Carinae) the second-brightest star in the whole sky after Sirius in Canis Major. Even farther south, and right on the horizon for people at 30°S (roughly the latitude of Sydney in Australia) is ***Achernar*** (α Eridani), the main star in the long, winding constellation of ***Eridanus*** (the River), which begins near the star Rigel in Orion, far to the north.

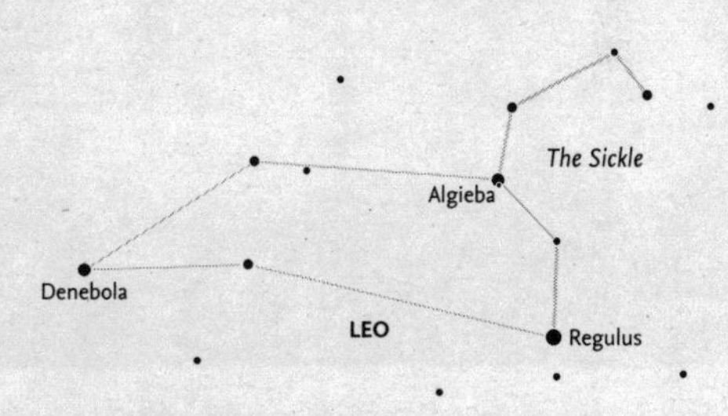

The distinctive constellation of Leo, with Regulus and 'The Sickle' on the west. Algieba (γ Leonis), north of Regulus, appearing double, is a multiple system of four stars.

The Moon Illusion

Frequently, when the Moon is seen rising or setting, and it is close to the horizon, it appears absolutely enormous. It seems far larger than when it is high in the sky. In fact, this is an optical illusion – it is known as the 'Moon Illusion'. Our eyes are playing tricks. In reality, the Moon is exactly the same size wherever it is in the sky. Try covering it with your finger, held at arm's length. Any finger is more than large enough to cover the Moon. (In fact, the Moon is about 30 minutes across, and a finger is about twice that, about 1 whole degree.) There have been many attempts over the years to explain the Moon Illusion, but it seems that the brain automatically compares the size of the Moon with the distant horizon, and assumes that it is at the same distance – whereas it is, of course, much farther away.

When the Moon is seen against distant objects, the brain is tricked into thinking it is much larger than its true size.

May

May – Introduction

The most significant astronomical event of May 2022 is the total lunar eclipse on May 16. This eclipse is visible from a wide area of the world. The whole event will be visible from South America. The maximum phases (when the Moon is darkest) will be visible from most of North America. Western Africa will see the Moon enter the umbra before the Moon sets, but will not see the central phase. The eclipse is described in more detail (with diagrams) on pages 114–115.

There is just one major meteor shower active in May. This is the ***η-Aquariids***, which are one of the two meteor streams associated with Comet 1P/Halley (the other being the Orionids, in October). The η-Aquariids are visible from both hemispheres, but are not particularly favourably placed for northern-hemisphere observers, because the radiant is close to the celestial equator, near the 'Water Jar' in Aquarius, and is well below the horizon until late in the night (around dawn). Meteors may still be seen in the eastern sky, however, even when the radiant is below the horizon. Their maximum in 2022, on May 6, occurs when the Moon is a waxing crescent (Day 6 of the lunation), so conditions are reasonably favourable. Maximum hourly rate is about 50 meteors per hour and a large proportion (about 25 per cent) of the meteors leave persistent trains.

The ***Pioneer Venus Orbiter*** mission was launched on 20 May 1978.

In May 2020, it was announced that two new southern meteor streams had been discovered. These streams are apparently associated with long-period comets. The first of these new meteor streams is the ***γ-Piscid Austrinids***, which appear to peak early in the morning of May 15 and may well be an annual shower. This date is one day before Full Moon in 2022, so observing conditions are particularly poor, and the shower will probably be poorly observed (if at all) this year. The second southern shower, (the ***σ-Phoenicids***) appeared to peak slightly later on the same date. The data suggest that the rate for this shower may vary considerably from year to year. Both showers are known for low numbers of meteors only (15 and 14, respectively) and have yet to be recognized as 'official' meteor showers.

The planets

Mercury is lost in daylight throughout May. No less than four planets are in the constellation of ***Pisces*** during the month. ***Venus*** (mag. -4.0) is the brightest, but rapidly moves closer to the Sun, becoming lost in morning twilight. There is an occultation of the planet on May 27, but this is essentially unobservable, occurring in daylight and visible only from a small area of the southern Indian Ocean. ***Mars*** (mag. 0.9 to 0.7) moves from ***Aquarius*** into Pisces in the morning sky. ***Jupiter*** (mag. -2.1 to -2.2), is also just inside Pisces. ***Saturn*** (mag. 0.8) is in ***Capricornus***. ***Uranus*** is in ***Aries*** at mag. 5.9 and comes to superior conjunction on May 6. ***Neptune***, at mag. 7.9, is initially close to the eastern border of ***Aquarius*** and, during the month, moves into Pisces.

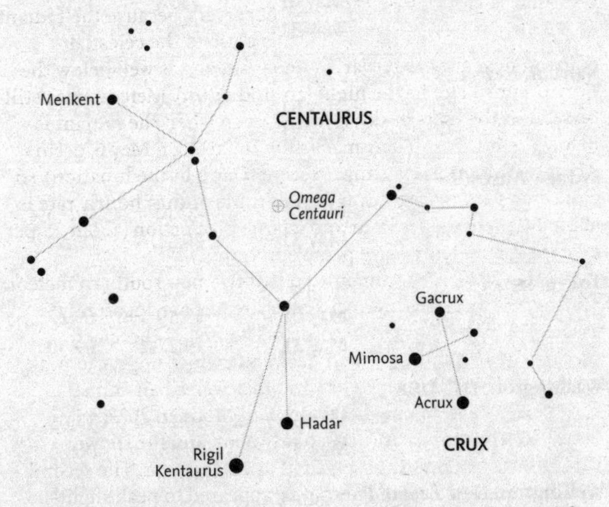

In May, the constellation of Centaurus is well placed for observing from the tropical regions and farther south. It contains the largest and finest globular star cluster in the sky, Omega Centauri. Crux, the Southern Cross is between the front and the hind legs of the Centaur.

Sunrise and sunset

City	Date	Sunrise	Sunset
Buenos Aires, Argentina			
	May 01	10:29	21:11
	May 31	10:51	20:51
Cape Town, South Africa			
	May 01	05:21	16:05
	May 31	05:42	15:45
London, UK			
	May 01	04:33	19:24
	May 31	03:50	20:08
Los Angeles, USA			
	May 01	13:04	02:37
	May 31	12:43	02:59
Nairobi, Kenya			
	May 01	03:28	15:32
	May 31	03:29	15:32
Sydney, Australia			
	May 01	20:30	07:15
	May 31	20:51	06:55
Tokyo, Japan			
	May 01	19:49	09:27
	May 31	19:27	09:50
Washington, DC, USA			
	May 01	10:10	17:05
	May 31	09:45	00:26
Wellington, New Zealand			
	May 01	19:09	05:28
	May 31	19:37	05:01

NB: the times given are in Universal Time (UT)

The Moon's phases and ages

Northern hemisphere

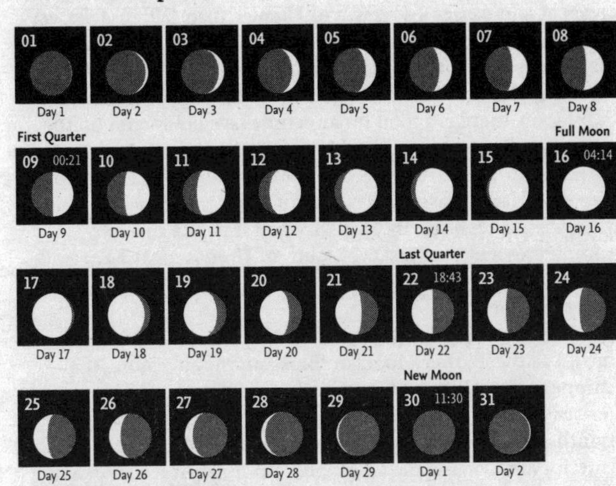

Southern hemisphere

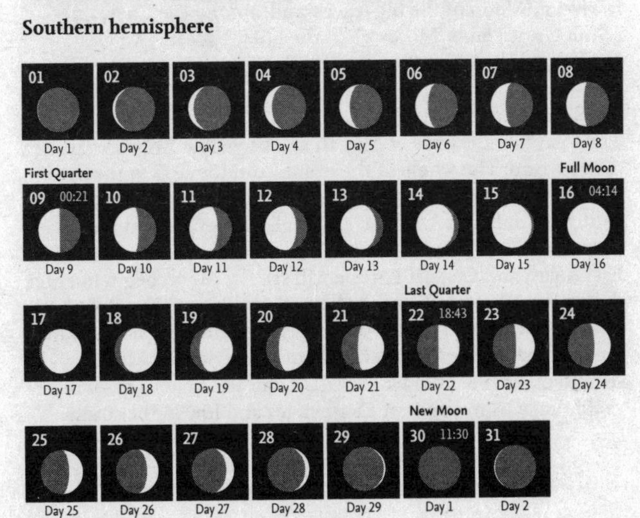

The Moon

On May 1, one day after New Moon, the Moon is a very thin crescent and passes 0.4° south of ***Uranus*** (mag. 5.9) in ***Aries***, very low in the morning sky. The Moon is 2.1° south of ***Pollux*** on May 6. It is 5.1° north of ***Regulus*** in ***Leo*** on May 9. On May 13, waxing gibbous, it is 5.1° north of ***Spica*** in ***Virgo***. On May 16, at Full Moon, there is a total lunar eclipse (see below and on the next page). The next day, the Moon is 3.1° north of ***Antares***. On May 22, at Last Quarter, it is 4.5° south of ***Saturn*** in ***Capricornus***. Two days later it passes 2.8° south of ***Mars*** and then 3.3° south of ***Jupiter*** in ***Aquarius***. By May 28, when it is just a narrow waning crescent, it is just 0.3° south of ***Uranus*** in ***Aries***.

Flower Moon

The Full Moon of May (this year on May 16) is sometimes known as the 'Flower Moon' or 'Blossom Moon' among the Chippewa and Ojibwe of the Great Lakes region. This term derives from, of course, so many flowers blooming during the month. Other names for this particular Full Moon from North America are 'Corn-Planting Moon', 'Moon when ice is breaking in rivers', 'Moon of the big leaves' and, among the Cheyenne of the Great Plains 'Moon when the horses get fat'. An Old English/Anglo-Saxon term was 'Milk Moon'.

Total lunar eclipse May 16

The total lunar eclipse of May 16 will be visible from a wide area of the world. The whole of the event from the first to the final contact with the Earth's ***penumbra*** will be observable from the continent of South America (although the penumbral phases are not detectable to the human eye). Mid-eclipse is at 04:11 UT. East Africa and Eastern Europe will see the Moon begin to enter the ***umbra***. Western Europe will see the whole of the total phase, but West Africa will see the Moon leave the umbra, very low in the sky, just before Moonset. In North America the Moon will be seen to enter the umbra at 02:27 UT, but the final stages will be visible only from the West Coast states and low in the south.

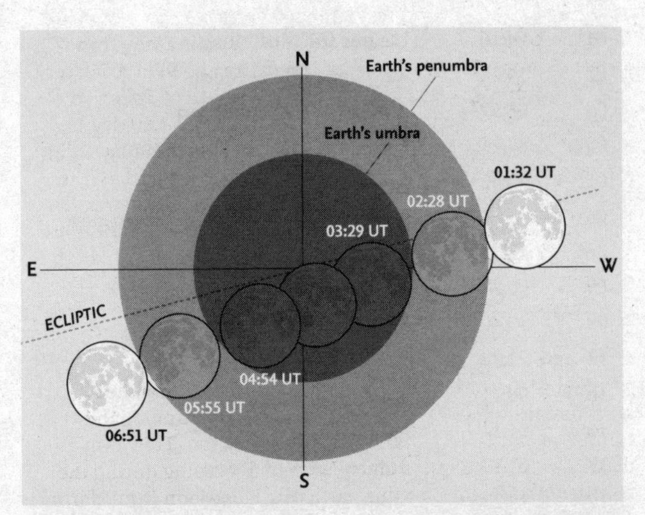

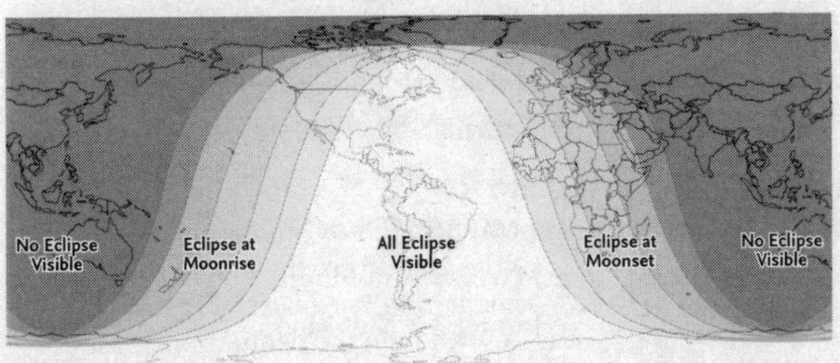

The lines on the map indicate where different phases of the eclipse occur. Starting in the east (right) they show when the Moon first contacts the penumbra; when it fully enters the penumbra; first contact with the dark umbra; and when it fully enters the umbra. The later lines indicate corresponding phases at the end of the eclipse: Moon leaving umbra; Moon fully outside umbra; Moon leaving penumbra.

Calendar for May

01	04:10	Uranus 0.4°N of Moon
02	14:18	Mercury 1.9°N of Moon
03	08:57	Aldebaran 7.2°S of Moon
05	07:21	Uranus at superior conjunction
05	12:46	Moon at apogee = 405,285 km
06		η-Aquariid meteor shower maximum
06	23:32	Pollux 2.1°N of Moon
09	00:21	First Quarter
09	19:40	Regulus 5.1°S of Moon
13	21:29	Spica 5.1S of Moon
16	04:11	Total lunar eclipse
16	04:14	Full Moon
17	03:19	Antares 3.1°S of Moon
17	15:27	Moon at perigee = 360,298 km
17	23:00 *	Neptune 0.6°N of Mars
21	19:18	Mercury at inferior conjunction
22	04:43	Saturn 4.5°N of Moon
22	18:43	Last Quarter
24	10:01	Neptune 3.7°N of Moon
24	19:23	Mars 2.8°N of Moon
25	00:02	Jupiter 3.3°N of Moon
27	02:51	Venus 0.2°N of Moon
28	13:42	Uranus 0.3°N of Moon
29	00:00 *	Mars 0.6°S of Jupiter
29	12:53	Mercury 3.7°S of Moon
30	11:30	New Moon
30	15:38	Aldebaran 7.2°S of Moon

** These objects are close together for an extended period around this time.*

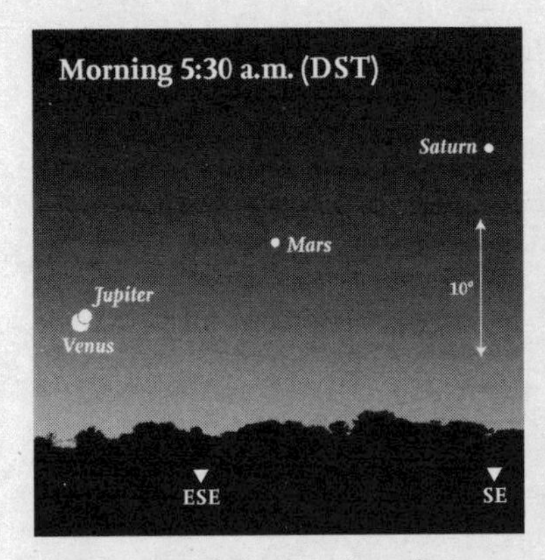

May 1 • *Four planets are almost lining up in the morning sky. Saturn, Mars, Jupiter and Venus (as seen from central USA).*

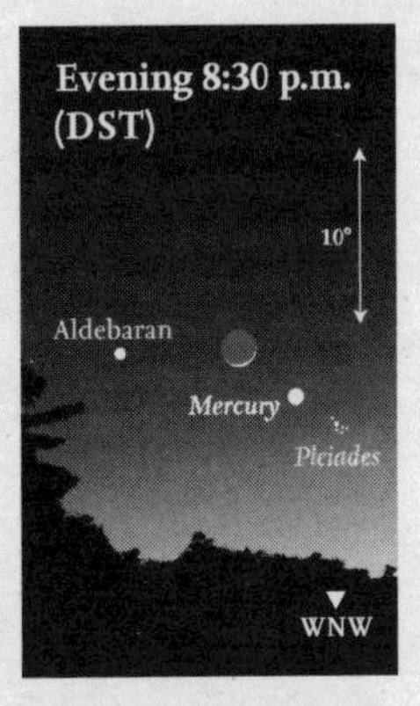

May 2 • *The Moon with Mercury, Aldebaran and the Pleiades (as seen from central USA).*

May 9 • *The Moon is between Regulus and Algieba (γ Leo), high in the west-southwest (as seen from London).*

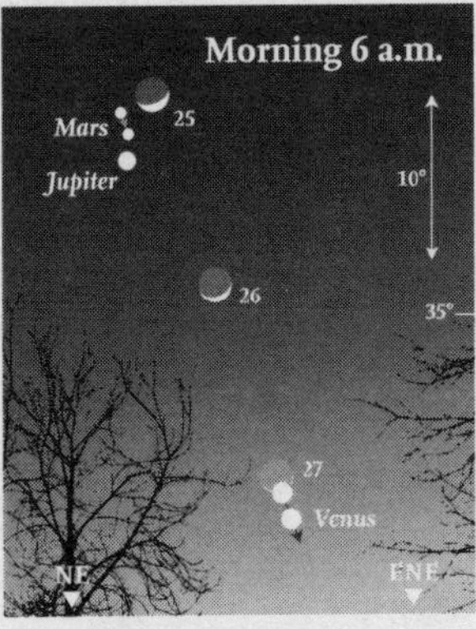

May 25–27 • *The waning crescent Moon passes Mars, Jupiter and Venus, in the northeast (as seen from Sydney).*

Early May 23:00 — Mid May 22:00 — Late May 21:00

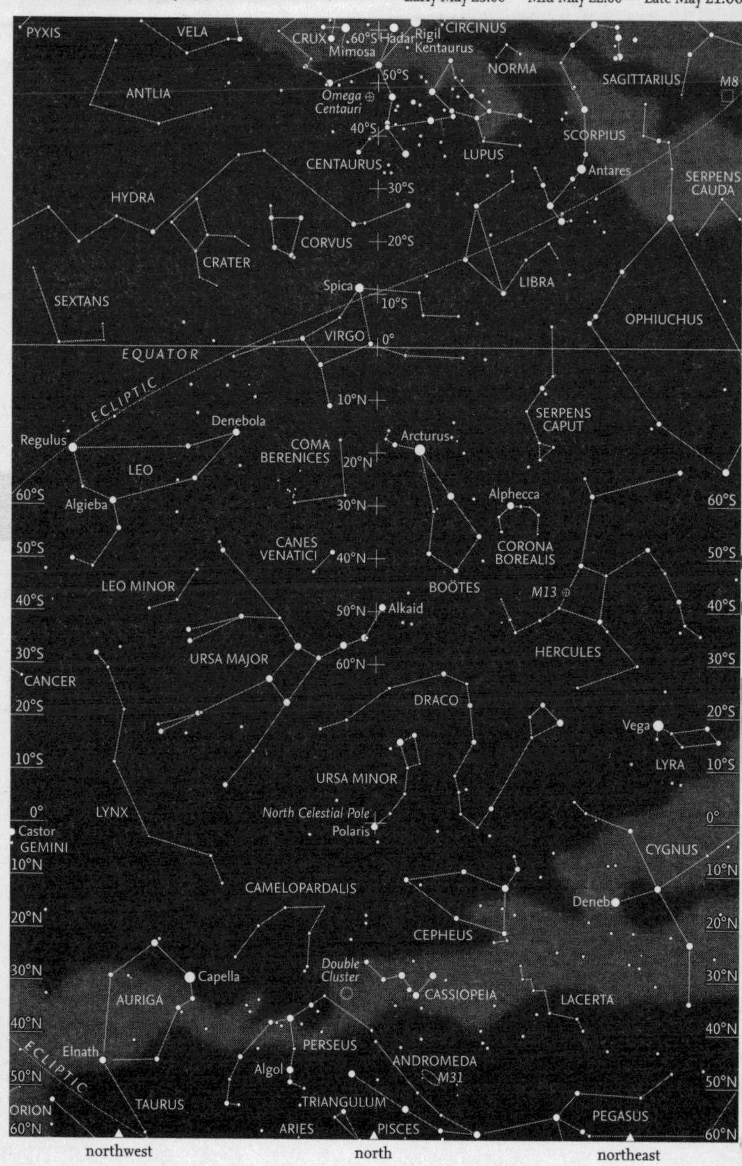

May – Looking North

The constellation of ***Cassiopeia*** is now low over the northern horizon, to the east of the meridian, to observers at mid-northern latitudes. To its west, the southern portions of both ***Perseus*** and ***Auriga*** are becoming difficult to observe as they become closer to the horizon, although the ***Double Cluster***, between Perseus and Cassiopeia, is still clearly visible.

High above, ***Ursa Major*** has started to swing round to the west and the stars that form the extended portion towards the south (the 'legs' and 'paws' of the 'bear') are becoming easier to see. ***Alkaid*** (η Ursae Majoris), at the end of the 'tail', is almost exactly at the zenith for observers at 50° North. The whole of the constellations of ***Cepheus***, ***Draco***, ***Lyra*** and ***Ursa Minor*** are easy to see, together with the inconspicuous constellation of ***Camelopardalis***. For those people with the keenest eyesight, they can even make out the long, straggling line of faint stars forming the constellation of ***Lynx*** in the northwestern sky, beyond the extended portion of Ursa Major. These faint constellations become difficult to detect during the brighter nights towards the end of the month.

The ***Mariner 9*** spaceprobe, designed to study Mars, was launched on 30 May 1971.

Also high in the sky for northern observers are the constellations of ***Boötes***, with bright ***Arcturus***, and ***Hercules***, with its clearly visible 'Keystone' of four stars, on one side of which is ***M13***, the finest globular cluster in the northern sky. The small, but striking arc of stars forming the constellation of ***Corona Borealis*** lies between Hercules and Boötes. It has just a single bright star, ***Alphecca*** (α Coronae Borealis).

Much of ***Cygnus*** is clearly seen in the east, as are two (***Deneb*** and ***Vega***) of the three stars that form the Summer Triangle. The third star, Altair in Aquila, begins to climb above the horizon late in the night and later in the month. Across the Milky Way, between Cassiopeia and Cygnus, and below Cepheus, is the tiny zig-zag of stars that forms the small, and often ignored, constellation of ***Lacerta***.

Early May 23:00 — Mid May 22:00 — Late May 21:00

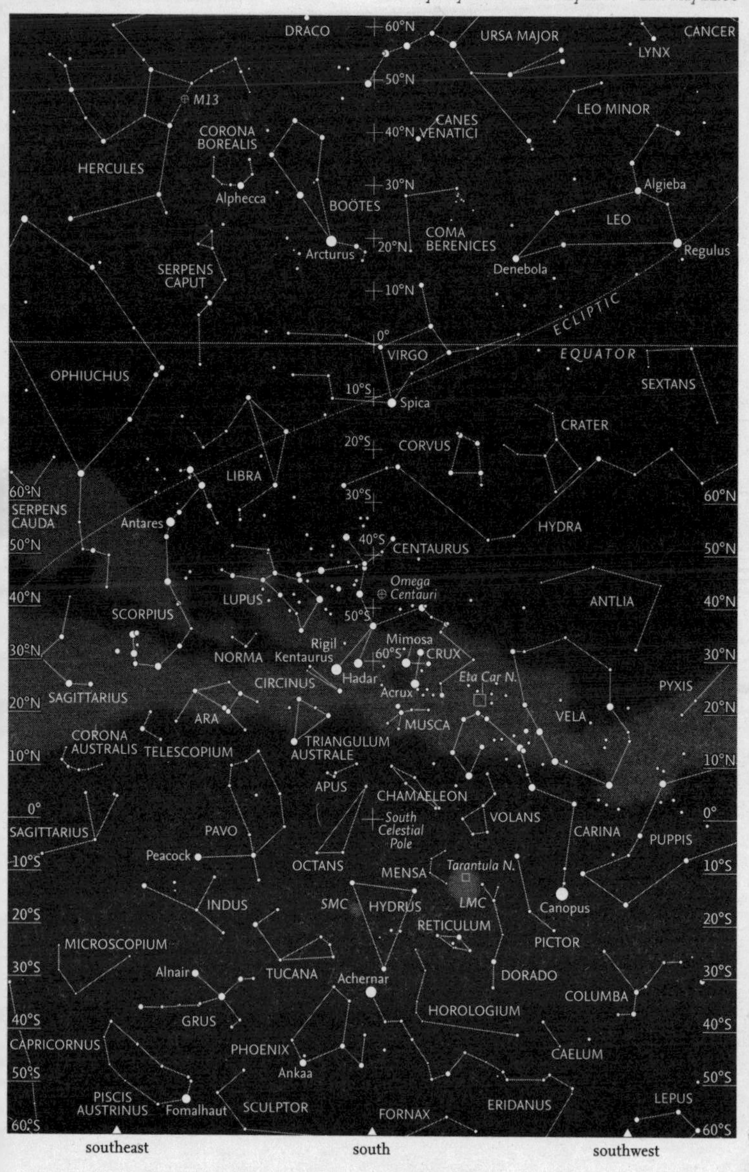

May – Looking South

Spica (α Virginis) in the sprawling constellation of ***Virgo*** is now close to the meridian. The constellation of ***Boötes***, with brilliant ***Arcturus*** (the brightest star in the northern hemisphere of the sky) is higher above and slightly to the east, clearly visible to observers in the north. Following Virgo across the sky is the rather undistinguished constellation of ***Libra*** and, following it, and rising in the east, ***Scorpius***, with the red supergiant star ***Antares*** and the distinctive line of stars, running south and ending in the 'sting'. (As Orion sinks in the west, so Scorpius rises in the east. This recalls one of the many legends about Orion: with his being pursued by a scorpion, sent to distract him while fighting.) Above Libra and Scorpius, the large constellation of ***Ophiuchus*** is beginning to climb higher in the sky. It lies between the two portions of ***Serpens*** (the only constellation to be divided into two parts). ***Serpens Caput*** (the Head of the Serpent) lies west of Ophiuchus, between it and Boötes, and ***Serpens Cauda*** (the Tail of the Serpent) is to the east, between Ophiuchus and the Milky Way.

Farther south, between Scorpius and the two brightest stars in ***Centaurus*** (***Rigil Kentaurus*** and ***Hadar***), lies the constellation of ***Lupus***, just east of the meridian and lying along the Milky Way. ***Crux*** is now more-or-less 'upright', with the small constellation of ***Musca*** below it. Below the bright pair of stars in Centaurus (α and β Centauri) is the constellation of ***Triangulum Australe***, a much larger and more striking constellation than its counterpart (Triangulum) in the north. Lying between Rigil Kentaurus and Triangulum Australe is the very tiny and indistinct constellation of ***Circinus***.

The stars of ***Vela*** and ***Carina*** are becoming lower in the southwest, following the constellation of ***Puppis*** and brilliant ***Canopus*** (α Carinae) down towards the horizon. Puppis and the ***Large Magellanic Cloud*** (LMC) are close to the horizon for anyone at the latitude of 30°S (about the latitude of Sydney in Australia), where ***Achernar*** (α Eridani) is now too low to be seen.

There are several, small, relatively faint constellations in this part of the sky. The most distinct is probably ***Pavo***, with its single bright star (α Pavonis), known as ***Peacock***. Other constellations are ***Apus***, ***Chamaeleon***, ***Octans*** (which actually includes the South Celestial Pole), ***Mensa*** and ***Volans***.

Eclipses

Solar eclipses

Between two and five solar eclipses may occur in any one year. During the period when dates have been reckoned on the basis of the Gregorian calendar (initially introduced in 1582), there have been six years in which there have been five solar eclipses. The last was 1935 and the next one will be 2206. There were four partial solar eclipses in 2011, and there will be two partial eclipses in 2022.

There are three (or four) types of solar eclipse. In a total eclipse, the whole of the Sun's disc is covered. This will normally occur when the Moon is close to perigee (closest to the Earth) and thus appears to subtend the greatest angle (diameter) from Earth. Such a situation occurs along a very narrow, central eclipse track across the surface of the Earth. The width of this track depends entirely upon the exact distance of the Moon from the Earth. When the Moon is farther from the Earth, particularly when it is at apogee (farthest away) it may not be large enough to cover the whole of the Sun's disc. We then have an annular eclipse, when, from certain locations a ring (an annulus) of the Sun's disc remains visible. Again, such annular eclipses are visible from a very restricted track on the surface of the Earth. Outside the central eclipse tracks, whether total or annular, only part of the Sun's disc is covered, giving a partial eclipse. A very rare form of eclipse is known as a hybrid eclipse (or an annular/total eclipse). In this, the eclipse is total for a very short duration in the central portion of its track and annular at the beginning and end of the event. This occurs because (on the spherical Earth) the two ends of the track are at different (greater) distances from the Moon than the central portion. The solar eclipse of 20 April 2023 will be a hybrid event.

On 5 May 2018, NASA's ***InSight*** spaceprobe landed successfully at Elysium Planitia on Mars and is currently operational.

The magnitude of any eclipse is the fraction of the Sun's disc that is covered by the Moon. For the two partial solar eclipses of 2022 (April 30 and October 25) the magnitudes are approximately 0.64 and 0.86, respectively (0.6389 and 0.8611, to be precise). Obviously, for a total eclipse, the magnitude will be 1 or greater. In annular eclipses, the magnitudes are obviously less than 1, and in the rare hybrid eclipses, the magnitude changes from being less than 1 at the beginning and end to slightly greater than 1 during the central phase.

Authorities disagree wildly on how much of the Sun must be covered for any change in illumination to be detected by normal human eyesight. The values quoted lie between 20 and 90 per cent. (Some of this wide range is due to differences in what is being discussed. In some cases it is the actual level of illumination, in others it is the percentage of the Sun's disc that has to be covered.) Certainly at least 70 per cent of the Sun needs to be covered for any alteration in the amount of sunlight to be readily detected. (A similar consideration applies to penumbral lunar eclipses, which are not detected by the human eye, and which only become visible when the Moon passes partially or completely within the umbra – the darkest part of the Earth's shadow.) If viewed with appropriate methods, partial solar eclipses are visible over fairly large areas of the Earth. The partial eclipse of 30 April 2022 will be visible over a wide region of the southeastern Pacific, off the western coast of South America (as shown in the diagram on page 98), and that of 25 October 2022 will be visible over a very large area of Asia, northeastern Africa, and most of Europe (page 193).

Lunar eclipses

In contrast to total and annular solar eclipses, total lunar eclipses are visible from a wide area of the Earth – in effect from anywhere the Moon is above the horizon. In any lunar eclipse, the Moon, as it moves eastward against the background stars, first enters the outer region of the Earth's shadow (the penumbra), where part of the Sun's disc is covered by the Earth. As it moves farther east, it encounters the region (the umbra) where the whole disc is hidden by the Earth. After the central phase, it will, of course, pass into the penumbra, and finally leave the shadow entirely. As mentioned, the diminution of light from the Moon is generally not detectable by the human eye until the Moon enters the umbra. The exact classification of the eclipse will depend on the proportion of the Moon that passes within the umbra.

In many cases the Moon's path does not take it fully within the umbra, giving a partial eclipse. Technically, even if a small section of the Moon remains within the penumbra, most being darkened within the umbra, the eclipse is still described as a partial eclipse. (The eclipse of 19 November 2021 was of this sort.) Both lunar eclipses in 2022 (of May 16 and November 8) are total, where the whole of the Moon passes into the central umbra (see pages 115 and 213). During the total phases, the colour of the Moon changes markedly. It is generally reddened, because red light penetrates through the ring (the annulus) of the Earth's atmosphere that an observer on the Moon would see around the Earth. The exact shade depends greatly upon the amount of cloud cover and may also be affected if there have been volcanic eruptions on Earth that have ejected material to great altitudes, where it tends to block the light, causing a dark eclipse. On very rare occasions, especially when it has passed through the centre of the shadow, the Moon has darkened to such an extent that it has even been said to have effectively disappeared. Frequently, the portion of the Moon that is covered by the edge of the shadow, i.e., close to the penumbra, appears a bluish shade, where some shorter-wavelength light passes through the Earth's outer atmosphere. Generally, some of the brighter features of the Moon, such as the craters Aristarchus, Proclus and Copernicus, remain detectable, even during the central phases of an eclipse.

June

June – Introduction

June is a very quiet month astronomically. On 21 June 2022, the Sun reaches its northernmost location, when it is directly overhead at the Tropic of Cancer, in the northern summer solstice. Because of the phenomenon of precession, the actual position of the Sun, at its extreme, at 09:14 on June 21, is no longer in Cancer, but in the constellation of Gemini.

Because of the light nights, when twilight persists throughout the night, astronomical observation is restricted and difficult for most northern observers, and the only compensation is that they may be able to observe noctilucent clouds, shining in the northern sky in the general direction of the North Pole. These clouds, which are the very highest in the atmosphere, occur towards the upper boundary of the layer known as the mesophere. This boundary, between the mesosphere and the next outermost layer, the exosphere, is known as the mesopause. Somewhat paradoxically, because of upwelling that occurs in summer, it lies at an altitude of about 100–120 km at that season, and its temperature is then lowest. It lies at about 80–85 km at other times of the year. The atmospheric temperature minimum occurs at the mesosphere and is rather variable, being usually between -163 and -100°C. Noctilucent clouds (NLC) consist of ice crystals and are generally found just below the mesopause. They are normally 'electric blue' in colour, although they tend to assume yellow/orange shades early in the night and towards dawn.

NLC are, of course, also visible in the southern hemisphere in the southern summer, but because of the lack of land masses in the south are less often seen. They are mainly reported by observers in the Antarctic Peninsula, although they have been seen from New Zealand, Tasmania, and from Patagonia in the southern tip of South America. In both hemispheres, NLC tend to be seen by observers located at latitudes between 50 and 60 degrees north or south.

Meteor observers, in particular, are at a distinct disadvantage this month. Throughout the whole month of June, no significant meteor streams are active. Activity resumes only in early July.

The planets

Mercury is at greatest western elongation (23.2°) in the evening sky on June 16, but too low to be readily visible. ***Venus*** (mag. -3.9), is initially on the borders of ***Pisces*** and ***Aries***, but swiftly moves closer to the Sun and becomes low in the evening sky. ***Mars*** (mag. 0.7 to 0.5) is moving eastwards in Pisces. ***Jupiter*** (mag . -2.3 to -2.4) is also in Pisces. ***Saturn*** (mag. 0.7 to 0.6) is still within ***Capricornus***, but begins retrograde motion on June 6. ***Uranus*** is slowly moving eastwards in Aries, brightening very slightly from mag. 5.9 to mag. 5.8. ***Neptune*** remains in Pisces at mag. 7.9.

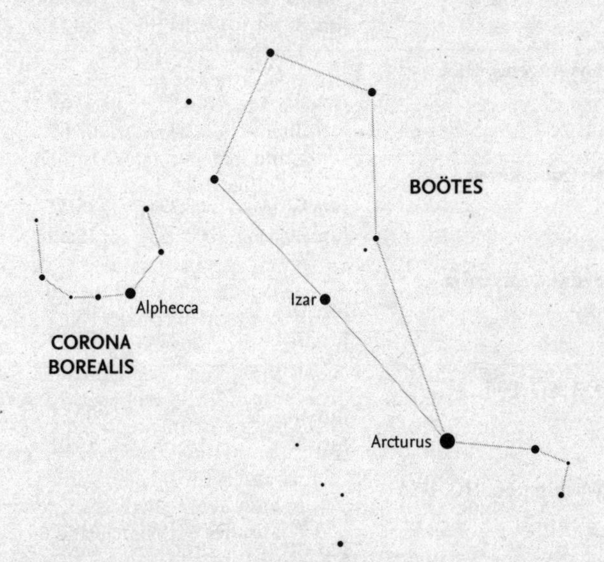

In June, the constellations of Boötes (with Arcturus) and Corona Borealis (with Alphecca) are well placed for observation. Arcturus has an orange tint and is the brightest star in the northern celestial hemisphere, at magnitude -0.05.

Sunrise and sunset

City	Date	Sunrise	Sunset
Buenos Aires, Argentina			
	Jun. 01	10:52	20:51
	Jun. 30	11:01	20:53
Cape Town, South Africa			
	Jun. 01	05:43	15:45
	Jun. 30	05:52	15:48
London, UK			
	Jun. 01	03:49	20:09
	Jun. 30	03:47	20:22
Los Angeles, USA			
	Jun. 01	12:43	02:59
	Jun. 30	12:45	03:09
Nairobi, Kenya			
	Jun. 01	03:29	15:32
	Jun. 30	03:35	15:38
Sydney, Australia			
	Jun. 01	20:52	06:54
	Jun. 30	21:01	06:57
Tokyo, Japan			
	Jun. 01	19:26	09:51
	Jun. 30	19:29	10:01
Washington, DC, USA			
	Jun. 01	09:45	00:27
	Jun. 30	09:46	00:38
Wellington, New Zealand			
	Jun. 01	19:38	05:00
	Jun. 30	19:48	05:01

NB: the times given are in Universal Time (UT)

The Moon's phases and ages

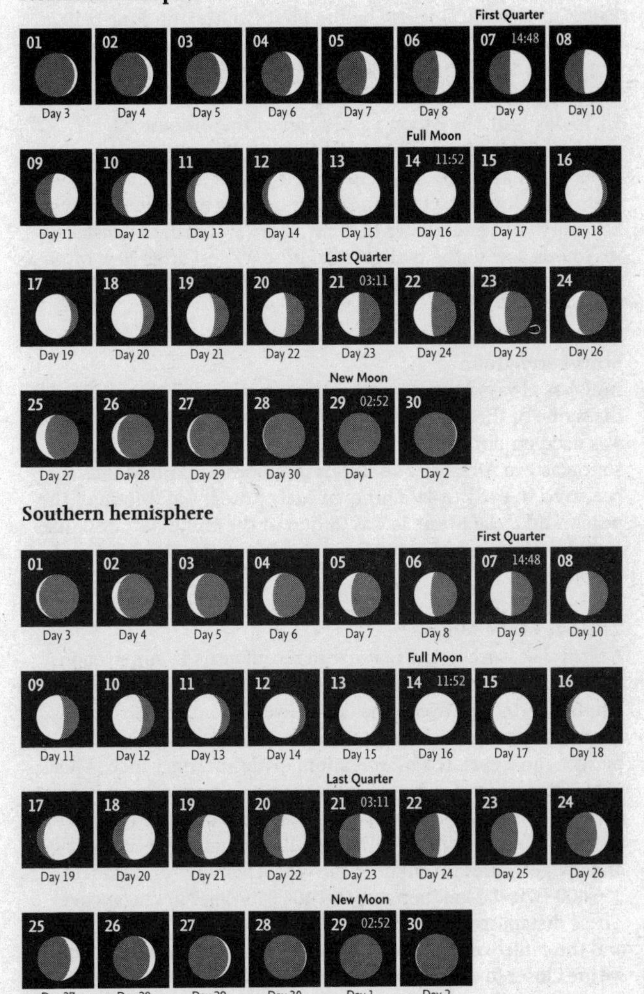

The Moon

On June 3, the waxing crescent Moon will be 2.1° south of ***Pollux*** in ***Gemini***. One day before First Quarter, on June 6, it will be 5.1° south of ***Regulus***. On June 10, it will be 5.0° north of ***Spica*** in ***Virgo***. On June 13, nearly Full, and in daylight, it passes 3.1° north of ***Antares***. Full Moon is on June 14. On June 18, in daylight, the Moon passes ***Saturn*** and on June 22, it is 2.7° south of ***Jupiter*** (again in daylight). It is 0.9° south of ***Mars*** the preceding day (June 21). As in May, it passes just south (0.1°) of ***Uranus*** on June 24. On June 26, in morning twilight, it passes 2.7° north of ***Venus***. Later that day, it is 7.2° north of ***Aldebaran*** in ***Taurus***, low in the twilight. On June 30, just after New Moon, as at the beginning of the month, it is again 2.2° south of Pollux.

Strawberry Moon

June has always been noted for strawberries, and this appears in the name of the Full Moon this month (on June 14). This name was used on both sides of the Atlantic. The Choctaw tribe of southeastern America had different names for Full Moons that occurred in early or late June. In early June it was 'Moon of the peach' and in late June is was 'Moon of the crane'. Other names are 'Hot Moon' and 'Rose Moon' and in the Old World, 'Mead Moon'.

The size of the Moon

A term that has become popular in recent years is 'supermoon'. The apparent size of the Moon varies naturally as a result of its elliptical orbit, relative to the Earth. Every month its distance alters from perigee (closest) to apogee (farthest). Two lunar perigees are separated by one 'anomalistic' month, which is just over 27.5 days (and slightly longer than one lunar orbit at just over 27.3 days). The dates of both extremes in the distance to the Moon are given in the monthly calendars. The actual distance at perigee and apogee varies throughout the year. The range of distances is 356,400–370,400 km (perigee) and 404,000–406,700 km (apogee). These distances are measured between the centre of the Earth and the centre of the Moon. Standing on the surface of the Earth, we are closer to the surface of the Moon by about 8000 km.

These changes in distance mean that the apparent size of the Moon alters continuously. However, the change in size is not readily apparent to the eye. Only photographs, taken at carefully arranged times, will show a difference when images at perigee and apogee are compared. When a Full Moon happens to occur at perigee, it is popularly termed a 'supermoon'. The display is not quite as dramatic as the name suggests, but such a Full Moon sometimes appears brighter than average. Another effect is seen when the Moon is close to the horizon, either rising or setting, and appears extremely large. This is known as the 'Moon Illusion', which is described on page 108. The Moon is not really larger, but the impression may be very striking.

These images of First and Last Quarter Moon (top) and Full Moon (bottom) show that the difference in size of the Moon at perigee and apogee is actually very small. (It is about 14 per cent, at most.) In both cases, the apogee image is on the right.

Calendar for June

02	01:13	Moon at apogee = 406,192 km
03	06:18	Pollux 2.1°N of Moon
06	03:07	Regulus 5.1°S of Moon
07	14:48	First Quarter
10	07:21	Spica 5.0°S of Moon
11	13:00 *	Uranus 1.6°N of Venus
13	13:58	Antares 3.1°S of Moon
14	11:52	Full Moon
14	23:23	Moon at perigee (closest of year) = 357,432 km
16	14:56	Mercury at greatest elongation (23.2°W, mag. 0.4)
18	12:22	Saturn 4.3°N of Moon
20	16:50	Neptune 3.5°N of Moon
21	03:11	Last Quarter
21	09:14	June solstice
21	13:36	Jupiter 2.7°N of Moon
22	18:16	Mars 0.9°N of Moon
24	22:13	Uranus 0.1°N of Moon
26	08:10	Venus 2.7°S of Moon
26	21:37	Aldebaran 7.2°S of Moon
27	08:18	Mercury 3.9°S of Moon
29	02:52	New Moon
29	06:08	Moon at apogee = 406,580 km
30	12:22	Pollux 2.2°N of Moon

* *These objects are close together for an extended period around this time.*

June 2 • *The crescent Moon with Castor and Pollux. Alhena (γ Gem) is close to the horizon (as seen from central USA).*

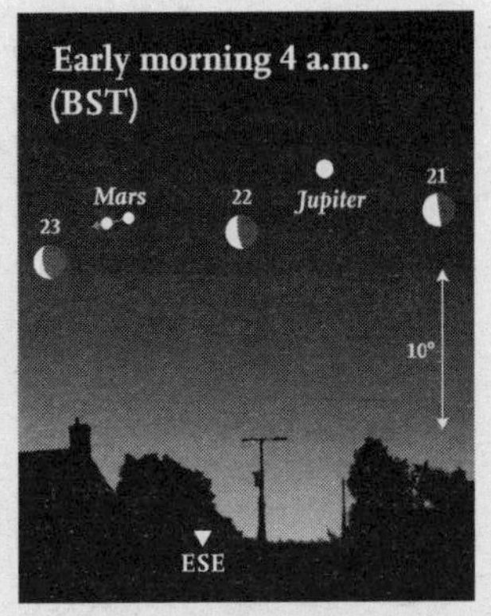

June 21–23 • *The Moon is at Last Quarter on June 21 and passes below Jupiter and Mars later (as seen from London).*

June 22–24 • *After Last Quarter, the Moon passes Jupiter and Mars (as seen from Sydney).*

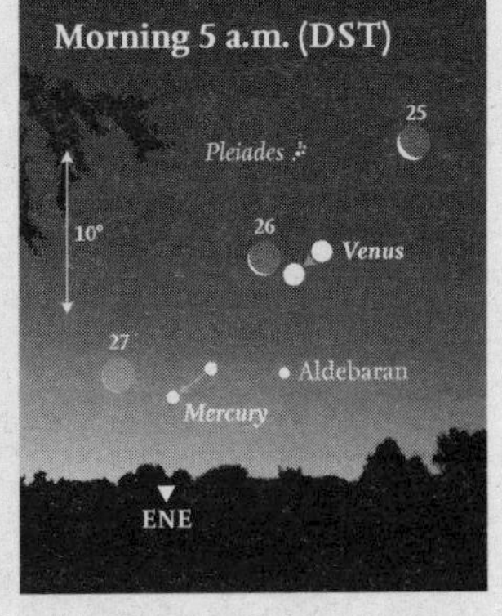

June 25–27 • *The Moon passes the Pleiades, Venus, Aldebaran and Mercury (as seen from central USA).*

Early June 23:00 — Mid June 22:00 — Late June 21:00

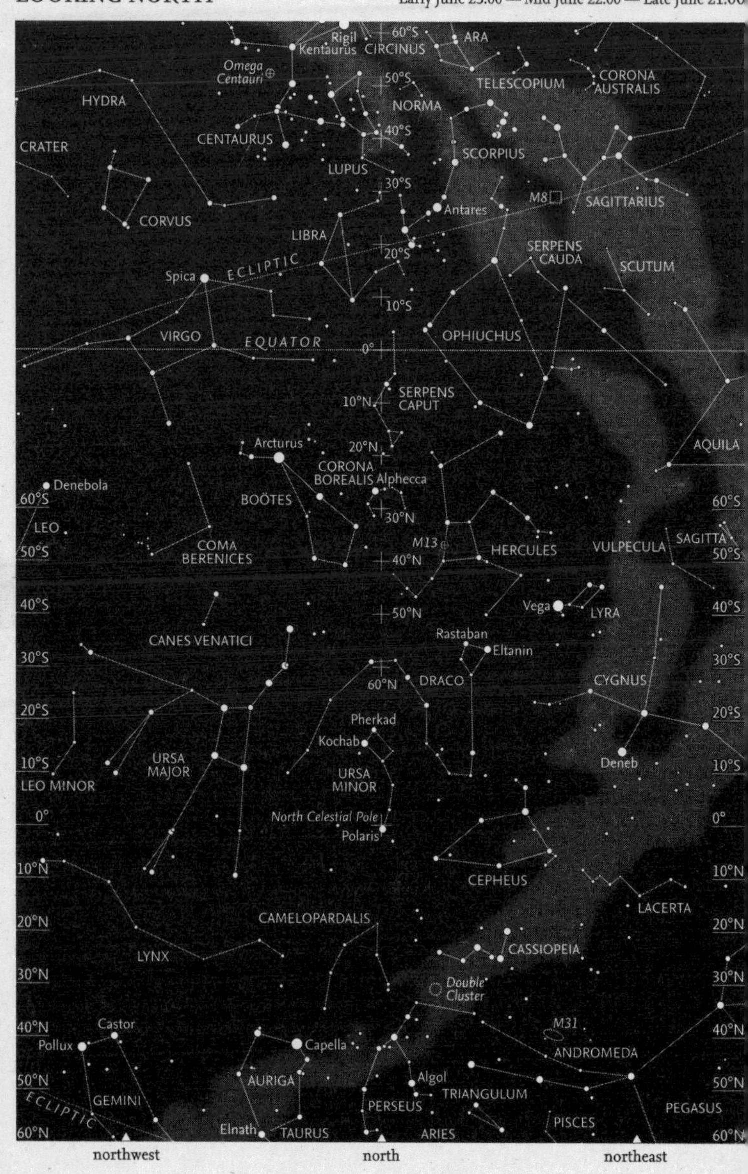

June – Looking North

For observers in the northern hemisphere, the time around the summer solstice (June 21) is frustrating for observing, because a form of twilight persists throughout the night. The sky remains so light that most faint stars and constellations are invisible unless conditions, such as the lack of light pollution, are particularly favourable. Even the highly distinctive set of seven stars forming the asterism of the 'Plough' (or 'Big Dipper') in ***Ursa Major*** may be difficult to see. The constellation has now swung round to the west, but the fainter stars of the constellation, lying even farther to the west, are harder to distinguish. The constellation of ***Boötes*** with bright ***Arcturus*** (α Boötis) is high overhead for northern observers, and best seen when facing south. The sprawling constellation of ***Hercules*** lies between ***Lyra*** and Boötes.

The four stars forming the 'head' of ***Draco*** lie east of the meridian, although only the brightest, ***Eltanin*** (γ Draconis) and, possibly ***Rastaban*** (β Draconis) are clearly visible. Farther north, in ***Ursa Minor***, are ***Polaris*** itself and the two 'Guards', ***Kochab*** (β UMi) and ***Pherkad*** (γ UMi).

For observers around 50°N, the southern portions of the constellation of ***Auriga*** are partially lost on the northern horizon, although bright ***Capella*** (α Aurigae) should still be visible. Much of the neighbouring, fainter constellation of ***Perseus*** is difficult to make out. The two main stars of ***Gemini***, ***Castor*** (α Geminorum) and ***Pollux*** (β Geminorum), the star closest to the ecliptic, and occasionally occulted by the Moon, are low on the northwestern horizon. Somewhat higher in the sky, and northeast of the meridian is ***Cassiopeia*** with ***Cepheus*** above it.

Two of the stars forming the angles of the distinctive Summer Triangle, ***Deneb*** (α Cygni) in ***Cygnus*** and ***Vega*** (α Lyrae) in Lyra are clearly visible as, for much of the night, is the third star, Altair (α Aquilae) in ***Aquila***. Deneb lies in the Milky Way, at the beginning of the dark Great Rift that runs down the constellation and where obscuring dust prevents us from seeing the dense star clouds of the Milky Way itself.

Early June 23:00 — Mid June 22:00 — Late June 21:00

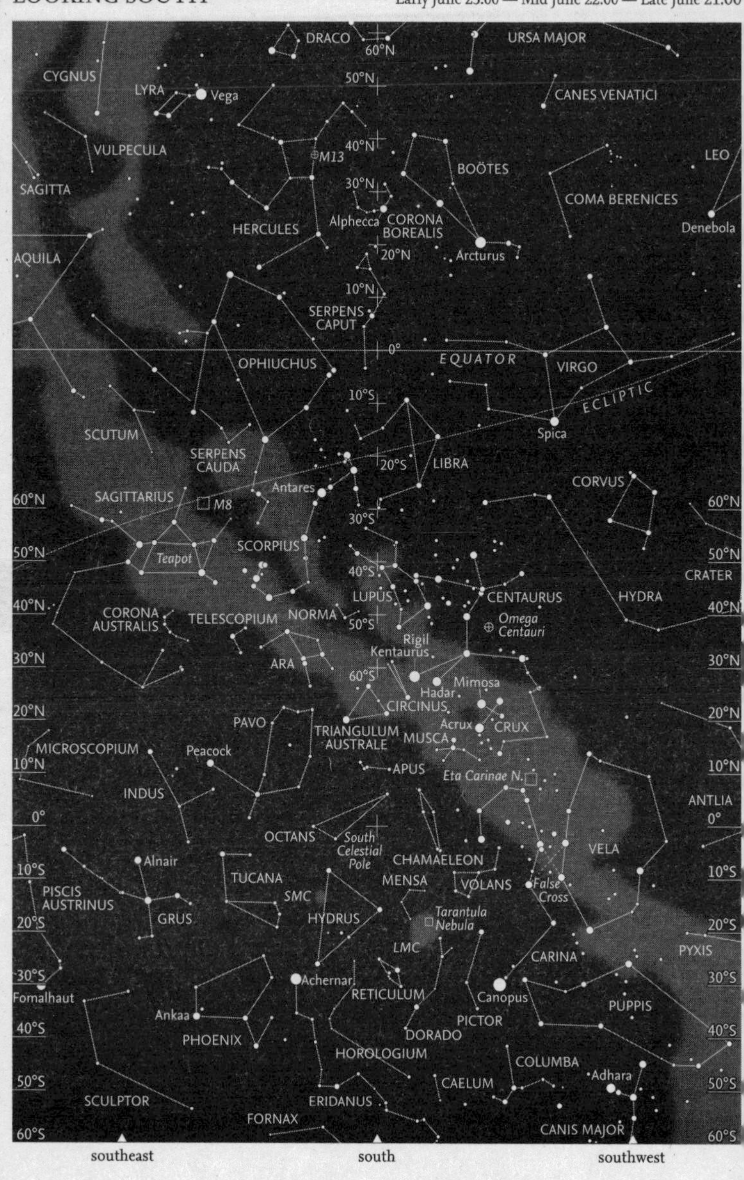

June – Looking South

The inconspicuous constellation of ***Libra*** is on the meridian, but the sky is really dominated by the striking constellation of ***Scorpius***, slightly to the east. These two were once a single constellation, of course, until the 'claws' of the scorpion were formed into the constellation of The Balance. Although it is rather low for most northern observers, to those farther south, fiery red ***Antares*** (α Scorpii) is a major beacon and the whole constellation has a very significant presence in the winter skies. The constellation of ***Virgo***, and bright ***Spica*** is now well to the west, along the ecliptic.

At this time of the year, the next constellation in the zodiac, ***Sagittarius***, is becoming clearly seen, having risen in the east. The main body of the constellation forms the asterism known as 'The Teapot', but the constellation also has a long, curving chain of faint stars to the south, rather similar to the long line of stars below Scorpius, although curving in the opposite direction. This chain of stars partially encloses the small constellation of ***Corona Australis***, although this constellation is not as well-formed as its northern counterpart (Corona Borealis). Rising in the east and following Sagittarius into the sky is the constellation of ***Capricornus***. To the west of the 'sting' of Scorpius is the rather untidy tangle of stars that forms the constellation of ***Lupus*** and part of ***Centaurus***.

The small constellation of ***Crux*** is now well south of the two brightest stars of Centaurus, ***Rigil Kentaurus*** and ***Hadar***. The constellations of ***Vela*** and ***Carina*** and the ***'False Cross'*** are plunging towards the horizon. For observers at the latitude of Sydney in Australia, both ***Canopus*** (α Carinae) and ***Achernar*** (α Eridani) are so low that they are often invisible through absorption at the horizon.

The South Celestial Pole in ***Octans*** is surrounded by faint constellations: Octans itself, ***Chamaeleon***, ***Volans***, ***Mensa***, ***Hydrus***, ***Tucana***, and ***Indus***. Only ***Pavo***, to the northeast and ***Grus***, to the southeast, are slightly more distinct. However, this area also includes the ***Large Magellanic Cloud*** (LMC), the largest satellite galaxy to our own. Most of the LMC lies in the constellation of ***Dorado***, although some extends into the neighbouring constellation of Mensa.

Comet Halley

The most famous comet of all, and one known to practically everyone, regardless of whether they take any interest in astronomy, is Comet Halley. Formally known as 1P/Halley, where the 'P' indicates that it is a periodic comet, and the '1' that it was the first to be properly identified, this comet has been known since ancient times. Returns of the comet have been recorded by astronomers since at least 240 BC, and possibly earlier. But it was not until AD 1696, that Edmond Halley,

'Looking at Halley's Comet' a watercolour by John James Chalon of the 1835 return (National Maritime Museum, Greenwich, London).

using his own observations of its return in 1682, together with observations of its returns in 1531 (observed by the astronomer Peter Apian) and 1607 (observed by Johannes Kepler), realized that it was periodic, and, using Newton's new theory of gravitation, was able to show that its orbit was elliptical, and that it would next return in 1758. It was recovered (after Halley's death) by Johan Georg Palitzsch on Christmas Day, 1758.

Modern-day calculations put the orbital period of Comet Halley at 75.32 years. However, because this period is less than 200 years, the comet is classed as a 'short-period' comet and is known to be affected by the gravitational effects of the planets, especially Jupiter. Attempts have been made to calculate exact return dates of early times, using accurate measurements made in the seventeenth and eighteenth centuries. However, these calculations could not be carried out farther back than AD 837, because of the unknown gravitation effects of a close encounter of the comet with the Earth in that year.

The ***Hayabusa 2*** spaceprobe went into orbit around the minor planet (162173) Ryugu on 27 June 2018.

The comet has been represented numerous times, the most famous being perhaps its representation of the 1066 apparition in the Bayeux Tapestry, where it was regarded as an ill omen, predicting the defeat and death of King Harold. Another famous representation is in the painting 'The Adoration of the Magi' by Giotto di Bondoni. Bondoni's painting was produced in about 1305, four years after the return of Comet Halley in 1301. This painter's name was the inspiration for the name of the European Space Agency's *Giotto* spaceprobe, which made the first ever close fly-by of any comet (which was Comet Halley) on 13–14 March 1986.

July

July – Introduction

Northern observers will continue to suffer from light nights during July, although there still remains the chance that they may be able to see noctilucent clouds during the first half of the month. (The noctilucent cloud 'season' in the north lasts about 6 to 8 weeks, centred on the summer solstice.)

The Earth reaches aphelion, the farthest point from the Sun in its yearly orbit, at 07:11 UT on July 4. Its distance is then 152, 098,455.06 km.

Meteor showers return in July. There are no less than four that are active. The first of these is the ***α-Capricornids***, which is visible from both northern and southern locations and does often produce bright fireballs. This is a moderately long shower that begins early in the month (about July 3) and continues until around August 15. Unfortunately, the maximum rate is very low, around 5 meteors per hour, the peak occurring on July 30. At the beginning of the shower and at maximum (such as it is) the Moon is a narrow waxing crescent, so those conditions are favourable. The end of shower, however, comes when the Moon is just past Full (which is on August 12). The parent body is a comet known as 169P/NEAT, which was one of the few comets discovered by NASA's Near-Earth Asteroid Tracking (NEAT) programme.

The Chinese ***Tianwen-1*** martian spaceprobe was launched from the Wenchang Spacecraft Launch Site on 23 July 2020.

The second shower active in July is the ***Southern δ-Aquariids***, also visible from both hemispheres. These begin in mid-month (on July 12, one day before Full Moon) and continue until August 23. Their maximum, like the α-Capricornids, is on July 30, with a waxing crescent Moon. Their hourly rate is somewhat higher than the α-Capricornids, but is still generally below 25 meteors per hour. The parent body is uncertain, but believed to be comet 96P/Macholz.

There is a minor southern shower, the ***Piscis Austrinids***, which is a moderately long shower, beginning about July 15 and continuing until August 10. Like the α-Capricornids, the maximum hourly rate is very low, around 5 meteors per hour, with maximum on July 28, at New Moon. The parent body is currently unknown.

By far the most important shower that begins in July is that of the ***Perseids***. These begin rather unfavourably around July 17, when the Moon is four days past Full, but have a strong maximum on August 12–13, so are more appropriately described next month.

The planets

Mercury is initially in the constellation of ***Taurus***, but rapidly moves towards the Sun. It comes to superior conjunction on the far side of the Sun on July 16. ***Venus*** is also initially in Taurus and is visible in the morning sky at mag. -3.9 to -3.8. ***Mars*** moves from ***Pisces*** into ***Aries***, brightening slightly from mag. 0.5 to 0.2. ***Jupiter*** (mag.-2.4 to -2.6) is generally located on the border of ***Pisces*** and ***Cetus***, and begins retrograde motion (that is, motion westwards on the sky) late in the month (on July 29). ***Saturn*** (mag. 0.6 to 0.4) continues its own retrograde motion in ***Capricornus***. ***Uranus*** is still in Aries at mag. 5.8, and ***Neptune*** is in ***Pisces*** at mag. 7.9.

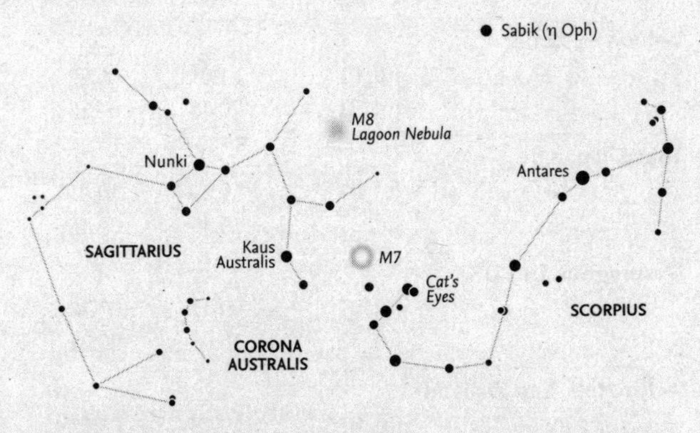

The sky looking south is dominated by the two zodiacal constellations of Scorpius and Sagittarius, on either side of the meridian, although they are low for observers at most northern latitudes.

Sunrise and sunset

City	Date	Sunrise	Sunset
Buenos Aires, Argentina			
	Jul. 01	11:01	20:54
	Jul. 31	10:48	21:12
Cape Town, South Africa			
	Jul. 01	05:52	15:48
	Jul. 31	05:40	16:06
London, UK			
	Jul. 01	03:47	20:22
	Jul. 31	04:23	19:51
Los Angeles, USA			
	Jul. 01	12:45	03:09
	Jul. 31	13:04	02:56
Nairobi, Kenya			
	Jul. 01	03:35	15:38
	Jul. 31	03:37	15:41
Sydney, Australia			
	Jul. 01	21:01	06:57
	Jul. 31	20:48	07:15
Tokyo, Japan			
	Jul. 01	19:29	10:01
	Jul. 31	19:49	09:46
Washington, DC, USA			
	Jul. 01	09:47	00:38
	Jul. 31	10:08	00:21
Wellington, New Zealand			
	Jul. 01	19:48	05:02
	Jul. 31	19:29	05:25

NB: the times given are in Universal Time (UT)

The Moon's phases and ages

Northern hemisphere

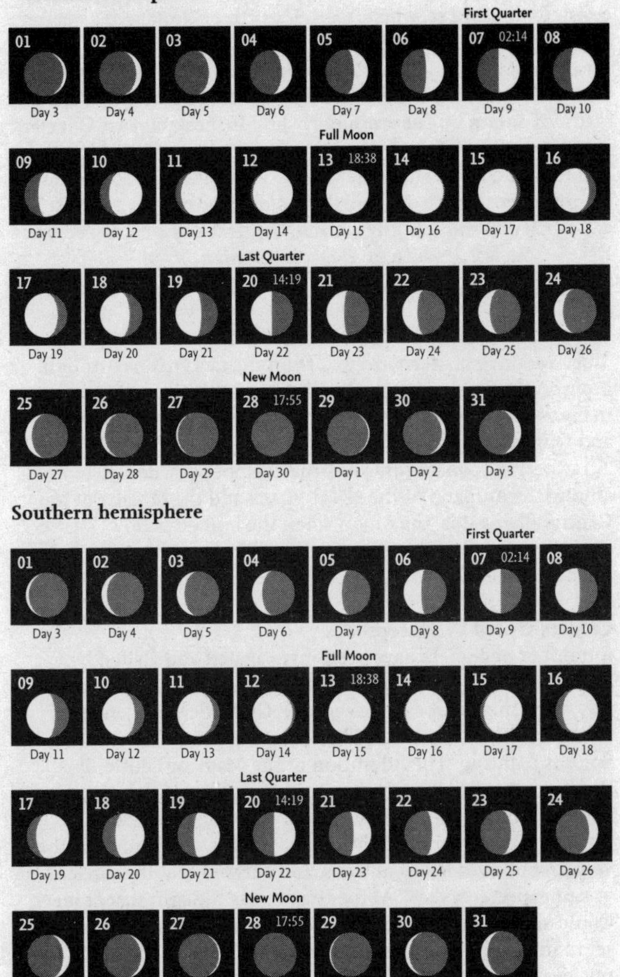

The Moon

On July 3, the Moon, a waxing crescent, passes through the constellation of ***Leo***, when it is 4.9° north of ***Regulus***, between it and ***Algieba***. On July 7 in daylight, it is the same distance (4.9°) north of ***Spica***. On July 11, two days before Full, it passes 3.0° north of ***Antares***. On July 15, two days after Full Moon, it is 4.1° south of ***Saturn*** in ***Capricornus***. By July 19, nearing Last Quarter, it is 2.2° south of ***Jupiter***. On July 21 it is 1.1° north of ***Mars***, and then, a day later, just 0.2° north of ***Uranus***. The Moon, now a waning crescent, is 7.4° north of ***Aldebaran*** on July 24. It is 4.2°north of ***Venus*** on July 26, and, the next day, one day before New Moon, is 2.2°south of ***Pollux*** in ***Gemini***.

Buck Moon

One of the names for the Full Moon for the month of July is 'Buck Moon'. This term derives from the fact that new antlers grow on the heads of male (buck) deer at this time. There are many other names for this Full Moon: among the Chippewa and Ojibwe of the Great Lakes area, following from the 'Strawberry Moon' of June, it is the 'Raspberry Moon'. Then, among the Arapaho of the Great Plains and the Omaha of the Central Plains it is 'the Moon when the buffalo bellow'. There are yet other terms from the Old English/Anglo-Saxon calendar, including the 'Thunder Moon', 'Wort Moon', and 'Hay Moon'.

Comets visited by spaceprobes

A number of comets have been investigated and visited by spaceprobes. The first close-up fly-by of a cometary nucleus was when the European spaceprobe ***Giotto*** flew by Comet Halley on 13 March 1986. (The probe was named after the famous painting 'The Adoration of the Magi' by Giotto di Bondoni, thought to contain a representation of the comet at its return in 1301.) The distance between the probe and the nucleus was 596 km, and the colour images obtained were the very first of a cometary nucleus, and showed that the nucleus was unexpectedly dark. At the time it was thought that comets would show large areas of exposed ices, and *Giotto's* imager was set to image the brightest part of the nucleus. In the event, the nucleus proved to be covered in dark, dusty particles, and this

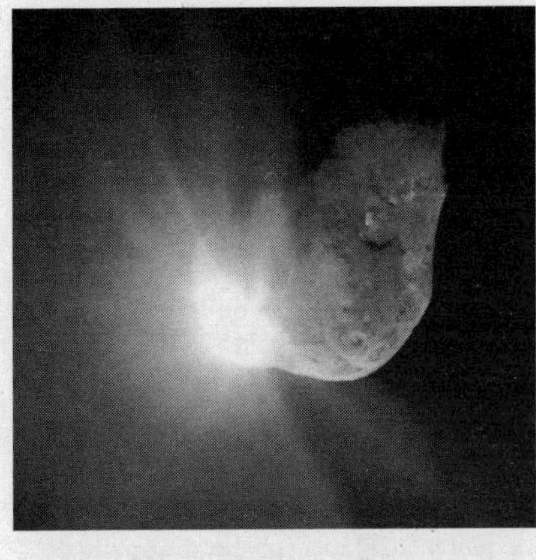

The brilliant cloud of dust particles created by the impactor from the Deep Impact *spaceprobe obscured the resulting crater and prevented imaging until the* Stardust *spaceprobe mission arrived nearly six years later.*

has been confirmed by later images of other comets. *Giotto* went on to fly past Comet ***26P/Grigg–Skjellerup***, in July 1992, at a distance of just 200 km, but was unable to return any images, because of damage sustained during the Halley encounter. Originally, the United States had intended to send a spaceprobe to accompany *Giotto*, but Congress imposed budget cuts at NASA, so that project was cancelled. Other spaceprobes were, however, reprogrammed to observe Comet Halley and provided valuable data that assisted with the navigation of the *Giotto* probe. The most notable of these were two Soviet spaceprobes, originally part of the ***Venera*** Venus-exploration programme. These were redesignated 'VeGa 1 and 2' and repurposed to observe both Venus and Comet Halley. They each released a landing probe and a small balloon orbiter on their arrival at Venus, ***Vega 1*** on 11 June 1985 and ***Vega 2*** on 15 June 1985. The *Vega 1* lander failed, but the *Vega 2* lander survived to reach the surface, from which it transmitted data for about 56 minutes. Both the balloon probes lasted for about two hours in the atmosphere of Venus.

The main *Vega* spacecraft were redirected, by using the gravity of Venus, to intercept the comet. This occurred the next year, *Vega 1* making its closest approach on 6 March 1986 at a distance of about 8890 km from the nucleus, and *Vega 2* following on 9 March 1985 at a closer distance of 8030 km. The two probes returned about 1500 images of the cometary nucleus.

The Japanese ***Suisei*** spaceprobe, originally known as PLANET-A, was designed to take images of the hydrogen corona of Comet Halley. It encountered the comet on 8 March 1986, passing at about 150,000 km from the sunwards side. Data from the Japanese ***Sakigake*** probe (carrying different instrumentation, but otherwise identical to *Suisei*) was used to refine the orbit of *Suisei*. *Sakigake* itself passed at a considerable distance (about 6.99 million km) from the comet.

The five spaceprobes: *Giotto*, *Vega 1*, *Vega 2*, *Suisei* and *Sakigake* are collectively known as the 'Halley Armada'.

The first attempt to investigate a comet was made by the ***International Cometary Explorer (ICE)*** spaceprobe, originally launched as the International Sun-Earth Explorer-3 (ISEE-3). (Two other, similar spaceprobes investigated the interactions between the magnetic fields of the Sun and Earth.) After being renamed and redirected, it became the first spaceprobe to visit any comet, passing through the plasma tail of comet ***21P/Giacobini-Zinner*** at about 7800 km from the nucleus.

Comet ***26P/Grigg–Skjellerup*** was indirectly studied, when it was discovered in 1972 that it was the source of the ☒-Puppid meteor shower (see pages 16 and 95). Analysis of particles collected by a high-flying NASA research aircraft in April 2003 revealed a previously unknown mineral, since named brownleeite, in honour of Donald Brownlee, an expert in cometary particles. During the comet's 1982 return, the Arecibo Observatory detected it by radar methods. The diameter of the nucleus is estimated to be 2.6 km.

The ***Stardust*** mission was launched in February 1999 and designed to capture particles from the coma of Comet Wild 2 (***81P/Wild***). This was the first sample return mission, and on its journey to the comet, the spaceprobe flew by the minor planet (5535) Annefrank. A sample capsule, containing cometary particles, was returned on 15 January 2005. The mission was extended and renamed NexT to fly by Comet Tempel 1, previously visited by the *Deep Impact* spacecraft. The spacecraft passed approximately 180 km from the nucleus on 15 February 2011, and successfully imaged the crater, estimated at 150 m in diameter, produced by the *Deep Impact* impactor.

On 21 September 2001, the ***Deep Space 1*** spacecraft performed a flyby of Comet 19P/Borrelly. Despite a systems failure, it was able to return images and scientific data about the comet.

On 12 January 2005, the ***Deep Impact*** spaceprobe was launched from Cape Canaveral. It encountered the comet Tempel 1 (***9P/Tempel***) and on 4 July 2005 fired an impactor at the nucleus. The impactor created a crater and also an unexpectedly large cloud of dusty debris, which obscured the crater and prevented the intended imaging. A subsequent flyby by the *Stardust* spaceprobe was able to image the resulting crater. The work at the comet produced the realisation that the surfaces of cometary nuclei generally consisted largely of dust particles and were less icy than previously expected.

Cometary science underwent a major advance with the investigation of Comet 67P/Churyumov-Gerasimenko by the *Rosetta* spaceprobe (pages 84–85).

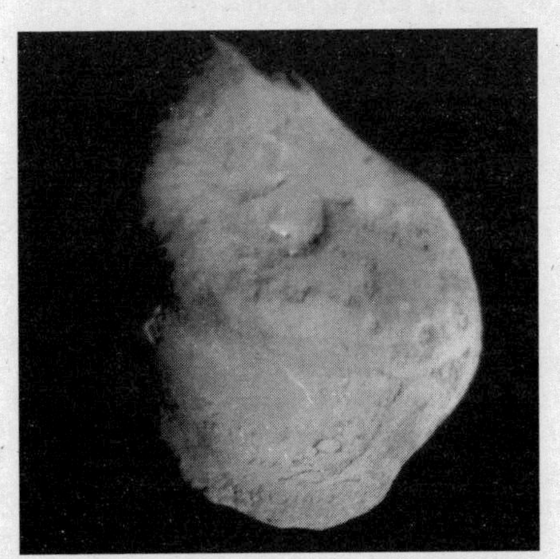

A composite image of Comet Tempel 1 as obtained by the impactor before it hit its target.

Calendar for July

02	00:00 *	Venus 4.2°N of Aldebaran
03–Aug.15		α-Capricornid meteor shower
03	09:22	Regulus 4.9°S of Moon
04	07:11	Earth at aphelion (1.016715 AU = 152,098,383 km)
07	02:14	First Quarter
07	15:37	Spica 4.9°S of Moon
11	00:22	Antares 3.0°S of Moon
12–Aug.23		Southern δ-Aquariid meteor shower
13	09:06	Moon at perigee = 357,264 km
13	18:38	Full Moon
15–Aug.10		Piscis Austrinid meteor shower
15	20:17	Saturn 4.1°N of Moon
16	19:38	Mercury at superior conjunction
17–Aug.24		Perseid meteor shower
18	00:50	Neptune 3.3°N of Moon
19	00:59	Jupiter 2.2°N of Moon
20	14:19	Last Quarter
21	16:46	Mars 1.1°S of Moon
22	06:21	Uranus 0.2°S of Moon
24	03:36	Aldebaran 7.4°S of Moon
26	10:22	Moon at apogee (most distant of year) = 406,274 km
26	14:11	Venus 4.2°S of Moon
27	18:21	Pollux 2.2°N of Moon
28		Piscis Austrinid meteor shower maximum
28	17:55	New Moon
29	21:07	Mercury 3.6°S of Moon
30		α-Capricornid meteor shower maximum
30		Southern δ-Aquariid meteor shower maximum
30	15:08	Regulus 4.8°S of Moon

* *These objects are close together for an extended period around this time.*

July 3–5 • *The waxing crescent Moon passes through the constellation of Leo, the Lion (as seen from Sydney).*

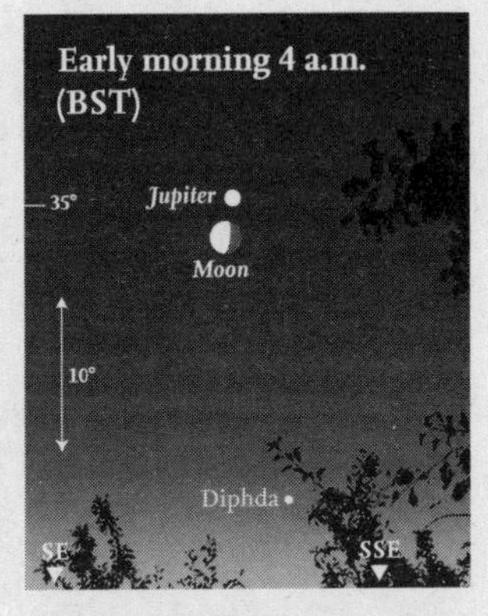

July 19 • *The Moon and Jupiter are close together, with Diphda (β Cet) closer to the horizon (as seen from London).*

July 25–27 • *The narrow crescent Moon passes Elnath (β Tau), Alhena (γ Gem), and Venus (as seen from Sydney).*

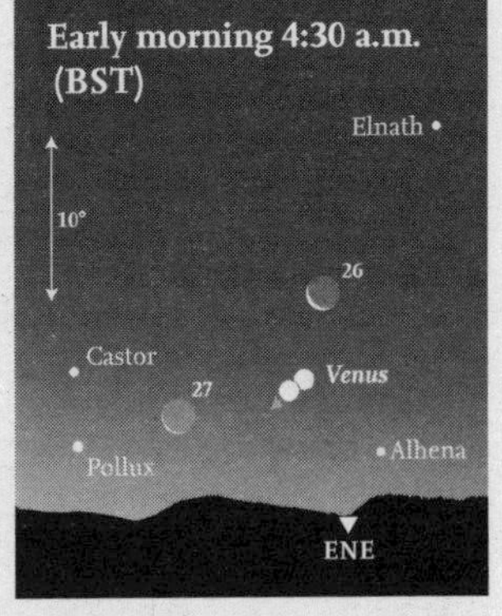

July 26–27 • *The Moon with Venus in the constellation of Gemini, the Twins (as seen from London).*

J

Early July 23:00 — Mid July 22:00 — Late July 21:00

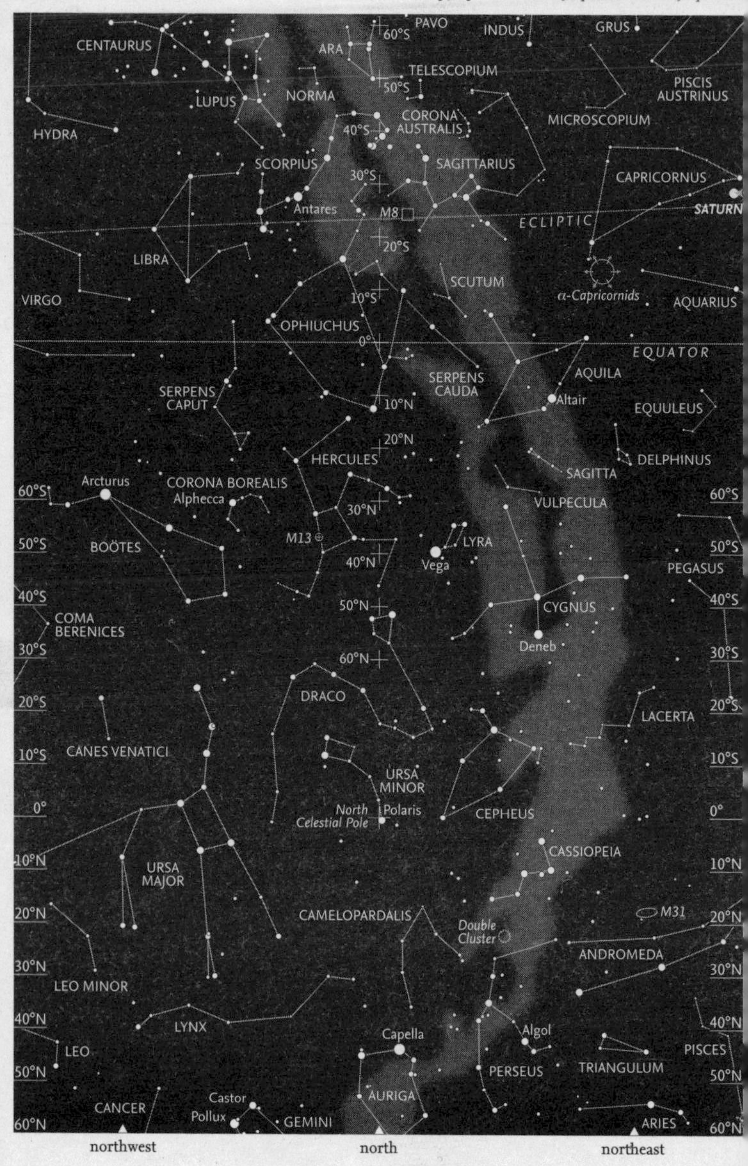

July – Looking North

The brilliant star ***Vega*** (α Lyrae) is now shining high overhead and the constellations of ***Hercules*** and ***Lyra*** are on opposite sides of the meridian, not far from the zenith for observers at 40°N, while it is the head of ***Draco*** that is near the zenith for observers slightly farther north at 50°N. Beyond Hercules, in the northwestern sky, are the constellations of ***Corona Borealis*** and ***Boötes***, the latter with brilliant, orange-tinted ***Arcturus***.

The stars of the Milky Way are now running more-or-less 'vertically', from north to south on the eastern side of the meridian. The constellation of ***Cygnus*** is 'upside down', high in the sky. For observers at the equator, it is the giant constellation of ***Ophiuchus*** that is at the zenith, with the two parts of ***Serpens*** (***Serpens Caput*** to the west, and ***Serpens Cauda*** to the east, among the clouds of the Milky Way). The third star of the Summer Triangle, ***Altair*** (α Aquilae) in ***Aquila*** is similarly high in the sky.

Ursa Major is now clearly visible in the northwest, and on the opposite side of the meridian, the constellation of ***Cepheus***, with its base in the Milky Way, is at a slightly greater altitude. ***Cassiopeia***, the other constellation that, like Ursa Major, is the key to finding one's way around the northern circumpolar constellations, lies in the Milky Way on the opposite side of the North Celestial Pole and ***Polaris*** in ***Ursa Minor***. The faint constellations of ***Camelopardalis*** and ***Lynx*** lie to the west and slightly farther south. The chain of faint stars forming Lynx runs 'horizontally' below the outflung stars of Ursa Major. Below Cassiopeia on the other side of the meridian, ***Perseus***, with the famous variable star, ***Algol***, is beginning to climb higher in the sky and observers at mid-northern latitudes will find that they can now more clearly see ***Capella*** (α Aurigae) and the northernmost portion of ***Auriga***. Observers in the far north (around latitude 60°N) may even occasionally glimpse ***Castor*** and ***Pollux*** in ***Gemini*** peeping above the northern horizon.

On 22 July 1972, the Soviet ***Venera 8*** probe was the second object to land successfully on the surface of Venus. It returned data for just over 50 minutes.

J

Early July 23:00 — Mid July 22:00 — Late July 21:00

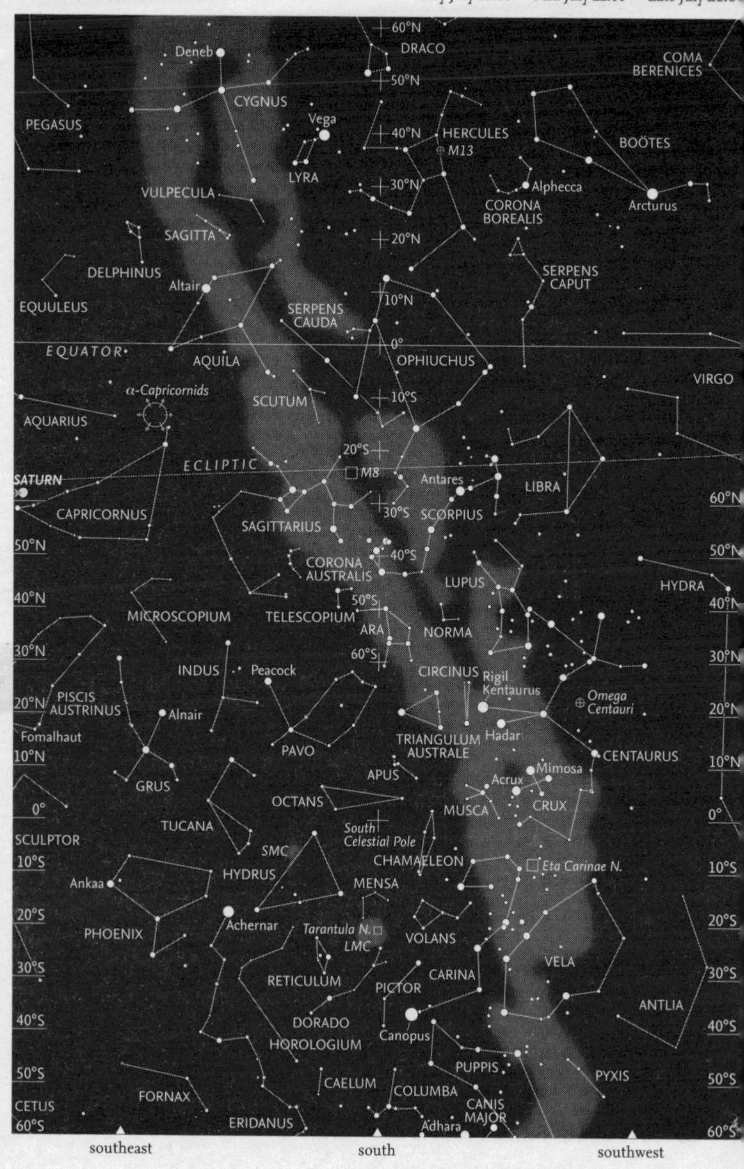

July – Looking South

The sky looking south is dominated by the two zodiacal constellations of ***Scorpius*** and ***Sagittarius***, which lie on the ecliptic on either side of the meridian, although they are low for observers at most northern latitudes. Bright, red ***Antares*** (α Scorpii) is very conspicuous, even when it is low in the sky. Not for nothing has it earned the name that means 'Rival of Mars'. The roughly triangular shape of ***Capricornus***, the next zodiacal constellation, lies east of Sagittarius and is now clearly visible. Below Scorpius, in the Milky Way, are the small, and often ignored, constellations of ***Norma*** (which is little more than three stars and easily overlooked) and ***Ara***, the latter with a more distinctive shape and brighter stars. The stars of ***Lupus*** and the outlying stars of ***Centaurus*** (including the great globular cluster known as ***Omega Centauri***) lie farther west.

High above, the constellation of ***Ophiuchus*** lies at the zenith for observers on the equator, with ***Aquila*** and bright ***Altair*** (α Aquilae) to the east. Northwest of Ophiuchus is ***Boötes***, the principal star of which is ***Arcturus***, which most people see as having an orange tint. Between Arcturus and the meridian is the circlet of stars that forms the constellation of ***Corona Borealis***.

South of Scorpius is ***Triangulum Australe***, to the east of the two bright stars of Centaurus, ***Rigil Kentaurus*** and ***Hadar***, and lying within the star clouds of the Milky Way. At about the same altitude is the constellation of ***Pavo*** and, beyond it, ***Indus***. Still farther east lies the elongated and somewhat distorted cross-shape that is the constellation of ***Grus***. ***Piscis Austrinus***, which lies south of Capricornus and Aquarius, has a single bright star, ***Fomalhaut*** (α Piscis Austrini). This ancient constellation (it was mentioned by Ptolemy, in the second century AD) has now risen in the east.

Crux and the adjoining constellation of ***Musca*** are now low on the horizon for observers at the equator, and the ***Small Magellanic Cloud*** (SMC) in ***Hydrus*** is actually below it. For observers farther south, ***Achernar*** (α Eridani) is now clearly visible, as is the constellation of ***Phoenix*** to the east. Only observers in the extreme south, however, will be able to see the whole of ***Vela*** and ***Carina*** as well as brilliant ***Canopus*** (α Carinae).

J

Following its successful return of samples from minor planet (162173) Ryugu in December 2020, spaceprobe ***Hayabusa 2*** has been targeted to fly by the body known as (98943) 2001 CC21 in July 2026 and rendezvous with another, (1998) KY26 in July 2031.

August

August – Introduction

Meteors

One of the best meteor showers of the year occurs in August. These are the famous ***Perseids***, which have long been well-known, even to the general public, partly because they are visible during the warm nights of summer. In some Catholic countries they are known as the 'Tears of Saint Lawrence' because they are visible on August 10, the date of his martyrdom. The Perseids are a long shower, generally beginning about July 17 and continuing until around August 24, with a maximum on August 12–13, when the rate may reach as high as 100 meteors per hour (and on rare occasions, even higher). In 2022, maximum is around Full Moon, so conditions are particularly poor. The Perseids are debris from Comet 109P/Swift-Tuttle (the Great Comet of 1862). Perseid meteors are fast and many of the brighter ones leave persistent trains. Some of the meteors during this shower also appear as bright fireballs.

There are three other meteor showers, which reach their peak maxima in July, but still exhibit activity that persists into August. These are the ***α-Capricornids***, which may be active until August 15 (just after Full Moon), the ***Southern δ-Aquariids*** (lasting until August 23) and the southern shower, the ***Piscis Austrinids***, which may persist until August 10. All these three showers have been described more fully on pages 14–16.

On 20 August 1977, the ***Voyager 2*** spaceprobe was launched. It studied Jupiter and Saturn, and made the first studies of Uranus and Neptune.

One short, minor, northern meteor shower, the ***α-Aurigids,*** begins late in the month, on August 28 in 2022, one day after New Moon, and reaches its weak peak, with 5 or 6 meteors per hour, on September 1.

Next page: *The path of Saturn in 2022, in the constellation of Capricornus. Saturn comes to opposition on August 14. Background stars are shown down to magnitude 6.5.*

The planets

The planet ***Mercury*** is very low in the twilight. It was at superior conjunction on July 16, but moves out from behind the Sun and comes to greatest elongation (27.3°) east at mag. 0.2, in the evening sky on August 27. ***Venus*** (mag. -3.8 to -3.9) is initially in ***Gemini***, but rapidly moves towards the Sun and into twilight. ***Mars*** moves from ***Aries*** (at mag. 0.2) into ***Taurus***, brightening to mag. -0.1.

On 25 August 2012, the ***Voyager 1*** spaceprobe crossed the heliopause and became the first object to enter interstellar space.

Jupiter, just within the northern region of ***Cetus***, continues its retrograde motion, slowly increasing from mag. -2.7 to mag. -2.9 over the month. ***Saturn*** is at opposition (mag. 0.3) in ***Capricornus*** on August 14. (A chart of the planet's position at opposition is shown below.) ***Uranus*** is still in ***Aries*** at mag. 5.8, and begins retrograde motion on August 25. ***Neptune*** is also retrograding slowly, moving from ***Pisces*** into ***Aquarius*** over the month, and is at mag. 7.9. On August 22, minor planet ***(4) Vesta*** is at opposition in Aquarius at mag. 5.8 (see charts on pages 172–173).

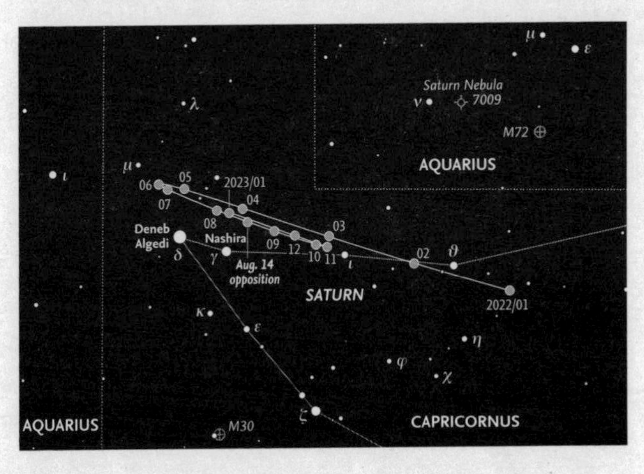

Sunrise and sunset

City	Date	Sunrise	Sunset
Buenos Aires, Argentina			
	Aug. 01	10:47	21:13
	Aug. 31	10:14	21:34
Cape Town, South Africa			
	Aug. 01	05:39	16:07
	Aug. 31	05:06	16:27
London, UK			
	Aug. 01	04:24	19:49
	Aug. 31	05:12	18:50
Los Angeles, USA			
	Aug. 01	13:05	02:55
	Aug. 31	13:26	02:21
Nairobi, Kenya			
	Aug. 01	03:37	15:41
	Aug. 31	03:31	15:36
Sydney, Australia			
	Aug. 01	20:47	07:15
	Aug. 31	20:14	07:36
Tokyo, Japan			
	Aug. 01	19:49	09:46
	Aug. 31	20:13	09:10
Washington, DC, USA			
	Aug. 01	10:09	00:20
	Aug. 31	10:36	23:40
Wellington, New Zealand			
	Aug. 01	19:28	05:26
	Aug. 31	18:47	05:55

NB: the times given are in Universal Time (UT)

The Moon's phases and ages

Northern hemisphere

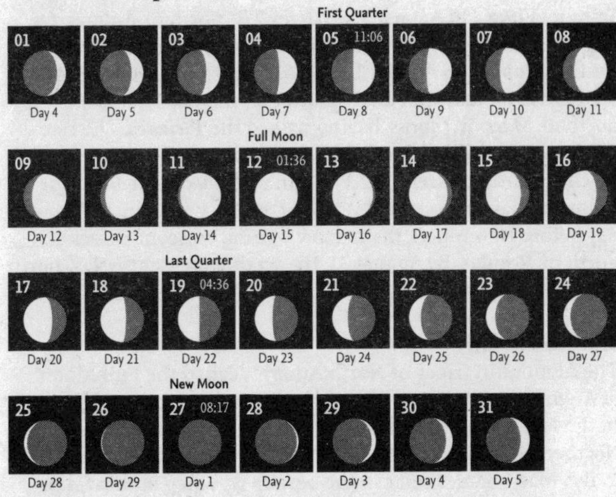

Southern hemisphere

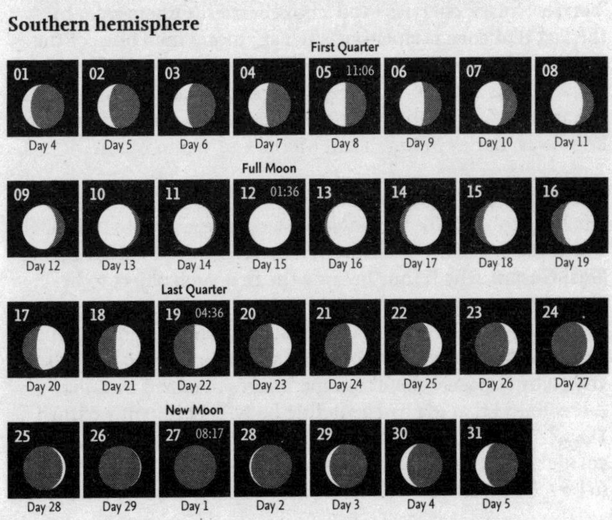

The Moon

On August 3, the waxing crescent Moon passes 4.6° north of ***Spica*** in ***Virgo***. On August 7, it passes 2.8° north of ***Antares***. On August 12, at Full Moon, it is 3.9° south of ***Saturn***. It passes 1.9° south of ***Jupiter*** on August 15 and 0.6° north of ***Uranus*** in ***Aries*** on August 18. Just after Last Quarter on August 19, it is 2.7° north of ***Mars*** in ***Taurus***, having passed the ***Pleiades***. The next day, it passes 7.6° north of ***Aldebaran***, still in Taurus. On August 24, the waning crescent is 2.1° south of ***Pollux*** in ***Gemini***. The following day it passes 4.3° north of ***Venus***. On August 26, one day before New Moon, the narrow waning crescent passes 4.7° north of ***Regulus***. By August 31, the waxing crescent is 4.4° north of ***Spica*** in ***Virgo***, as it was at the beginning of the month.

Sturgeon Moon

The Algonquin tribes of North America called the Full Moon of August the 'Sturgeon Moon', because of the numerous fish in the lakes where they fished. To the Cree of the Canadian Northern Plains it was the 'Moon when young ducks begin to fly'. Many names refer to the berries ripening at this time: 'berries'; 'black cherries'; and 'chokeberries'. Others refer to the fact that corn is ripening, such as among the Ponca of the Southern Plains: 'Corn is in the silk Moon'. To the Haida in Alaska it was the 'Moon for cedar bark for hats and baskets'. In the ancient Old English/Anglo-Saxon calendar it was sometimes known as 'Barley Moon', 'Fruit Moon' and 'Grain Moon'.

Minor planets visited by spaceprobes

On 29 October 1991, the *Galileo* spaceprobe, en route to Jupiter, which it reached in December 1995, flew past the minor planet ***(961) Gaspra***, which thus became the first such object to be viewed from a spaceprobe.

On 28 August 1993, the Galileo spaceprobe passed the minor planet ***(243) Ida***, making it only the second minor planet to be visited by a spaceprobe. From the images obtained by *Galileo* it was found that Ida had a satellite body, subsequently named ***Dactyl***. Ida itself has a diameter of 31.4 km, but Dactyl is far smaller, being only about 1.4 km across. The first minor planet to be fully investigated from orbit by a spaceprobe was ***(433) Eros***,

Minor planet	Spaceprobe	Year	Mission
307 Nike	*Pioneer 10*	1972	distant flyby
951 Gaspra	*Galileo*	1991	flyby
243 Ida	*Galileo*	1993	flyby
253 Mathilde	*NEAR- Shoemaker*	1997	flyby
433 Eros	*NEAR- Shoemaker*	1998–2001	orbited/landed
9989 Braille	*Deep Space 1*	1999	flyby
2685 Masursky	*Cassini-Huygens*	2000	flyby
25143 Itokawa	*Hayabusa*	2005	sample return
132524 APL	*New Horizons*	2006	distant flyby
5535 Annefrank	*Stardust*	2008	flyby
2867 Šteins	*Rosetta*	2008	flyby
21 Lutetia	*Rosetta*	2010	flyby
4 Vesta	*Dawn*	2011–2012	orbited
4179 Toutatis	*Chang'e-2*	2012	flyby
162173 Ryugu	*Hayabusa 2*	2018–2019	sample return
486958 Arrokoth	*New Horizons*	2019	flyby
101955 Bennu	*OSIRIS-REx*	2020	sample return

In addition to the objects shown in this table, on its travel to the outer Solar System, before reaching Jupiter, the Pioneer 10 *spaceprobe flew past an unnamed minor planet in 1972.*

which was the first near-Earth object to be discovered (in 1898) and which later proved to be the second-largest. Parallax measurements using Eros were made to determine the value of the astronomical unit (AU) at the minor planet's opposition in 1900–1901 and later at the close-approach in 1930–1931. These measurements provided the most accurate value of the astronomical unit until superseded by radar measurements in the late 1960s.

The minor planet Ida. A mosaic of images obtained by the Galileo *spaceprobe, just before its closest approach.*

A

It was visited in 1998 by the spaceprobe *NEAR-Shoemaker* and eventually in 2000, the spaceprobe's directors managed to place it in orbit around Eros. It actually landed on the body in the following year, on 12 February 2001. Eros was found to be a peanut-shaped (or potato-shaped) object, some 34 km long, with an average diameter of 16.8 km.

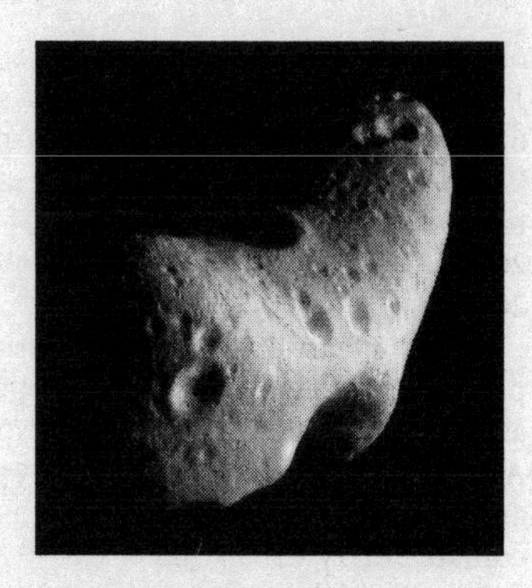

A mosaic of images of the minor planet (433) Eros as obtained by the NEAR-Shoemaker *spaceprobe.*

In 2005 the Japanese ***Hayabusa*** mission visited the small S-type minor planet ***(25143) Itokawa***, which is about 535 metres long and has a mean diameter of about 330 metres. The aim was to obtain a sample for return to Earth. The mission was only partly successful, because instead of a large surface sample, the probe succeeded in returning just a few small particles.

In 2011 the ***Dawn*** spaceprobe was inserted into orbit around the large minor planet ***(4) Vesta.*** It remained in orbit, carrying out extensive mapping and scientific observations until 2012 (see page 172), when it was sent on the dwarf planet Ceres.

On 27 June 2018, the Japanese spaceprobe ***Hayabusa 2*** reached and entered orbit around the minor planet ***(162173) Ryugu.*** This is a primitive, carbonaceous-type body (see pages 74–75). It has an effective diameter of approximately 850 metres, but has been found to be effectively a 'rubble-pile' with approximately 50 per cent empty space in its interior. The spaceprobe transported four small rovers to carry out measurements on the surface. Three functioned fully, and the fourth suffered a malfunction before release, but was released to orbit Ryugu and carry out gravitational and other measurements. On 5 April 2019, the spaceprobe fired a projectile at the surface to create a crater. It collected a sample of the material from the sub-surface exposed by this crater on 11 July 2019. *Hayabusa 2* left Ryugu in November 2019 and on 6 December 2020 successfully returned its sample container to Woomera in Western Australia. The spaceprobe has been sent towards new targets. It should fly by the body known as (98943) 2001 CC_{21} in July 2026 and rendezvous with (1998) KY_{26} in July 2031.

On 1 January 2019, the ***New Horizons*** spaceprobe flew past the distant, Kuiper-Belt object then temporarily, known as ***(486958) Ultima Thule*** and subsequently named ***(486958) Arrokoth.*** This is the farthest object yet visited by any spacecraft and was then 43.3 AU from the Sun. The body proves to be a contact binary, consisting of two distinct lobes, the larger and flatter of which appears to itself consist of several (perhaps 8?) different pieces that have aggregated together before the two lobes came into contact.

Before its encounter with (486958) Arrokoth, *New Horizons* had flown past the dwarf planet ***Pluto*** and its major satellite ***Charon*** on 14 July 2015. Following its encounter with Arrokoth, the spaceprobe may be targeted at another Kuiper-Belt object.

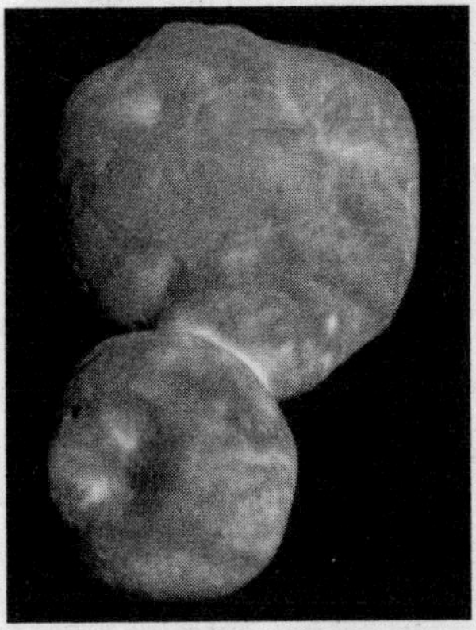

An image of Arrokoth, which is the most distant object surveyed by any spaceprobe.

On 3 December 2018, the spaceprobe *OSIRIS-REx* reached the minor planet ***(101955) Bennu.*** On 20 October 2020, it touched down and collected a sample of this primitive, carbonaceous minor planet. This object is smaller than the minor planet (162173) Ryugu visited by *Hayabusa* 2, with a maximum diameter of just 565 metres. The spaceprobe is expected to return the sample to Earth on 24 September 2023.

Calendar for August

01	09:00 *	Uranus 1.4°N of Mars
03	22:01	Spica 4.6°S of Moon
05	11:06	First Quarter
07	09:02	Antares 2.8°S of Moon
07	10:00 *	Venus 6.6°S of Pollux
10	17:09	Moon at perigee = 359,828 km
12–13		Perseid meteor shower maximum
12	01:36	Full Moon
12	03:56	Saturn 3.9°N of Moon
14	09:53	Neptune 3.1°N of Moon
14	17:10	Saturn at opposition (mag. 0.3)
15	09:42	Jupiter 1.9°N of Moon
18	14:38	Uranus 0.6°S of Moon
19	04:36	Last Quarter
19	12:17	Mars 2.7°S of Moon
20	10:24	Aldebaran 7.6°S of Moon
22	18:55	Minor planet (4) Vesta at opposition (mag. 5.8)
22	21:52	Moon at apogee = 405,418 km
24	00:53	Pollux 2.1°N of Moon
25	20:57	Venus 4.3°S of Moon
26	21:25	Regulus 4.7°S of Moon
27	08:17	New Moon
27	16:14	Mercury at greatest elongation (27.3°E, mag. 0.2)
28–Sep.05		α-Aurigid meteor shower
29	10:51	Mercury 6.6°S of Moon
31	03:35	Spica 4.4°S of Moon

* *These objects are close together for an extended period around this time.*

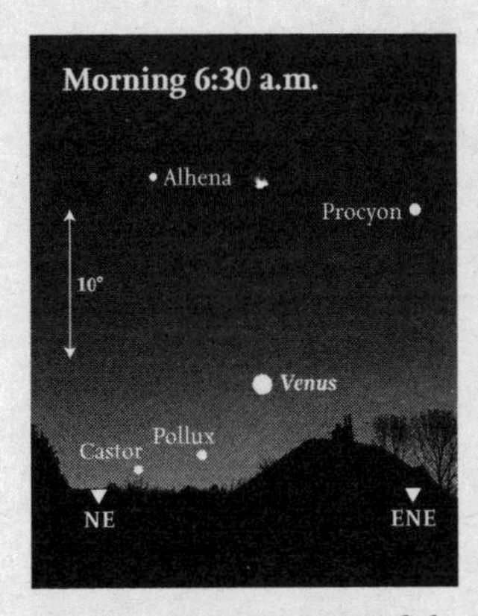

August 7 • *Venus is surrounded by Procyon, Alhena, and close to the horizon, Castor and Pollux (as seen from Sydney).*

August 15 • *The Moon and Jupiter are close together, with Diphda about 20 degrees lower (as seen from central USA).*

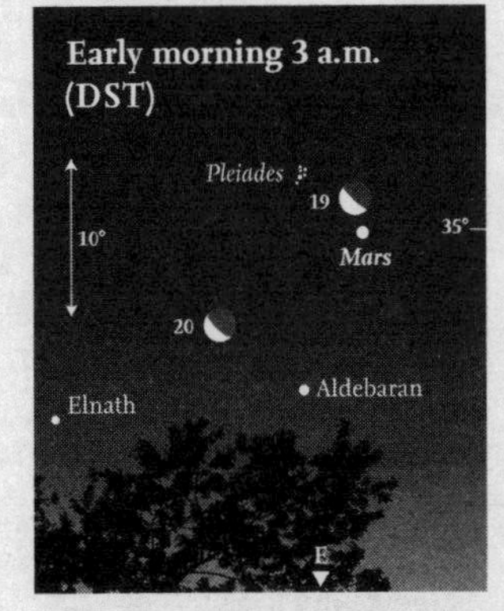

August 19–20 • *The Moon passes Mars, the Pleiades and Aldebaran (as seen from central USA).*

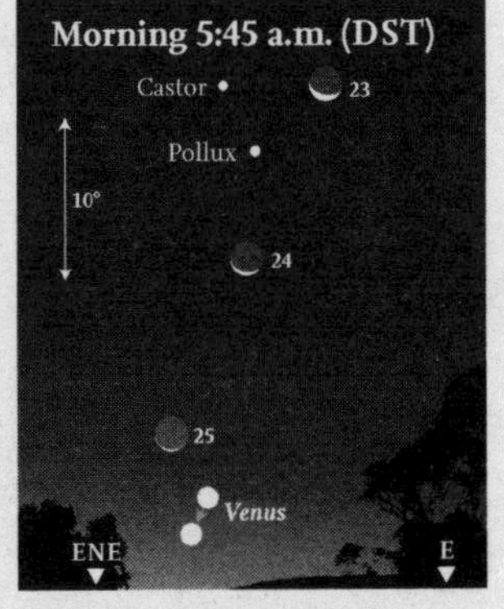

August 23–25 • *The Moon passes below Castor and Pollux. On August 25 it is close to Venus (as seen from central USA).*

A

Early August 23:00 — Mid August 22:00 — Late August 21:0

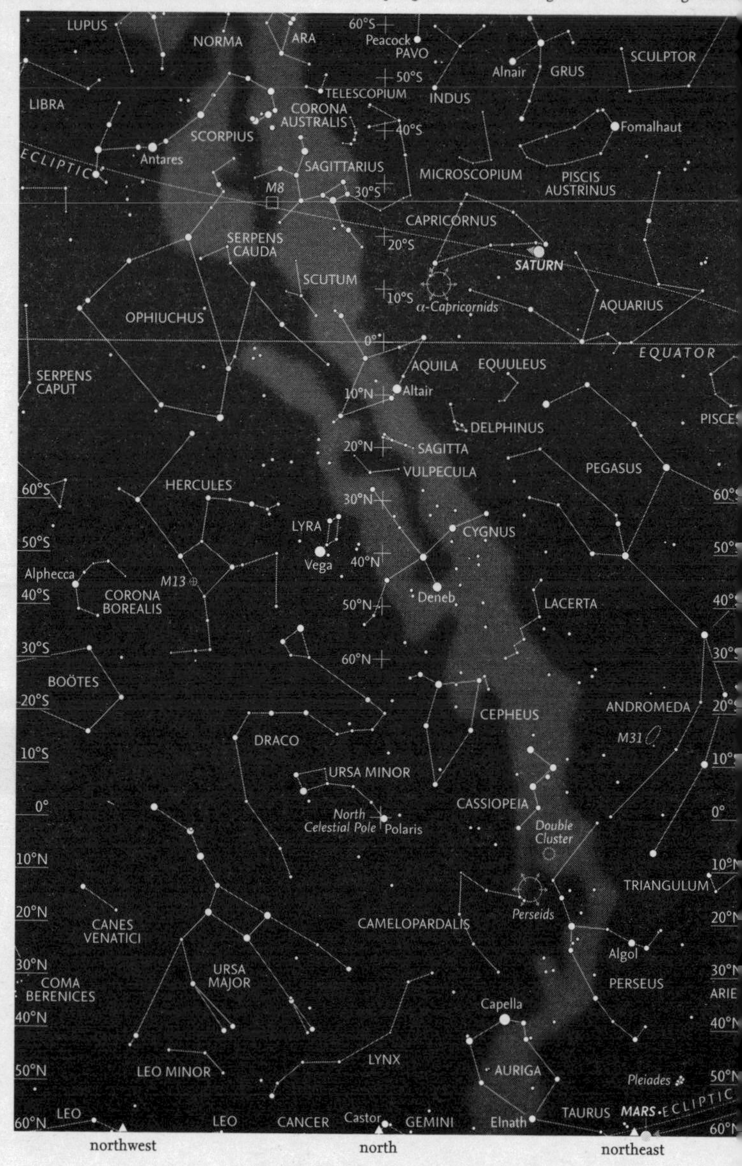

August – Looking North

The Milky Way is now running up the eastern side of the sky, where ***Cassiopeia*** may be found within it, high in the northeast. The constellation of ***Cepheus***, with the 'base' of the constellation, which is shaped like the gable-end of a house, lies within the edge of the Milky Way, and the whole constellation is 'upside-down' slightly farther north than Cassiopeia. For observers south of the equator, these constellations are close to the horizon and ***Ursa Minor*** is on the horizon.

For most mid-latitude northern observers, ***Perseus*** is now clearly visible, as is the northern portion of ***Auriga*** with brilliant ***Capella*** (α Aurigae). Perseus straddles a narrow portion of the Milky Way and although not particularly distinct, a significant section of the Milky Way actually runs through Auriga in the direction of ***Gemini***. Auriga contains numerous open clusters and a few emission nebulae, but is without globular clusters. Beyond Perseus in the east lies ***Andromeda*** and the small constellation of ***Triangulum***. Slightly farther north than Perseus and on the other side of the meridian is ***Ursa Major***, and for most observers even the stars in the southern, extended portion are visible.

For observers at about 40°N, both ***Lyra*** and ***Cygnus***, with their brilliant principal stars, ***Vega*** and ***Deneb***, respectively, are high overhead, near the zenith. The third star marking the other apex of the Summer Triangle, ***Altair*** in ***Aquila***, is much farther to the south, and all three stars are best seen when looking south. The constellation of ***Hercules*** lies farther west of Lyra and Cygnus, with most of the constellation of ***Pegasus*** to the east. The tiny, but highly distinctive constellation of ***Delphinus*** lies between the Milky Way and the outlying stars of Pegasus. Again, these areas are best seen when looking south.

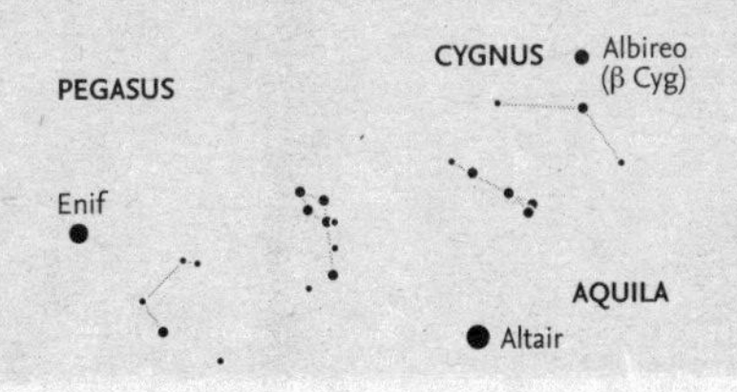

There are four small constellations between Cygnus, Pegasus, and Aquila. They are, from left to right: Equuleus, Delphinus, Sagitta, and Vulpecula.

A

Early August 23:00 — Mid August 22:00 — Late August 21:00

August – Looking South

The star clouds of the Milky Way dominate the sky at this time of the year. The three stars forming the Summer Triangle, ***Deneb*** (α Cygni), ***Vega*** (α Lyrae), and ***Altair*** (α Aquilae) are clearly visible, as is the Great Rift, running south from near Deneb. The dark Great Rift marks the location of dense clouds of dust that obscure the light from the many stars in the main plane of the Galaxy. Both ***Cygnus*** and ***Aquila*** represent birds, and in modern charts these are shown flying 'down' the Milky Way, towards the south. Older charts tended to show Aquila as flying 'across' the Milky Way, in the direction of ***Aquarius***, in the east.

For observers at northern latitudes, the constellations of ***Hercules*** and ***Pegasus*** are visible, one on each side of the meridian, together with ***Delphinus*** on the east. Between Cygnus and Aquila, in the Milky Way, lie the two small constellations of ***Vulpecula*** and ***Sagitta***.

The two zodiacal constellations of ***Scorpius*** and ***Sagittarius*** are still clearly visible, although becoming rather low for observers at mid-northern latitudes, with even the bright red supergiant star of ***Antares*** in Scorpius becoming difficult to see. (For observers at 50°N, it is skimming the horizon.) North of Scorpius is the large constellation of ***Ophiuchus***, with part of ***Serpens***, ***Serpens Cauda***, lying in front of the Great Rift.

Capricornus is clearly seen, as is ***Aquarius***, the next constellation along the ecliptic. For observers farther south, the curl of stars forming ***Corona Australis*** lies south of Sagittarius and, farther east, ***Piscis Austrinus*** with its single bright star, ***Fomalhaut*** (α Piscis Austrini) is at about the same altitude, with the chain of faint stars that curves south from Sagittarius and the faint constellation of ***Microscopium*** in between the two. South of Piscis Austrinus is the constellation of ***Grus***, with the undistinguished constellation of ***Indus*** between it and ***Pavo***, which is on the meridian.

Farther south, ***Lupus***, ***Centaurus*** and ***Crux*** are descending towards the horizon, while for observers at 30°S, ***Vela*** has disappeared, as has most of ***Carina***, including brilliant ***Canopus***.

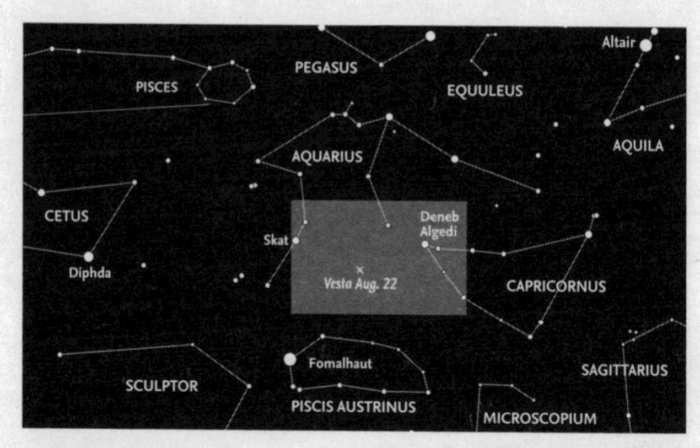

A finder chart for the position of minor planet (4) Vesta at its opposition on August 22. The grey area is shown in more detail on the next page.

The southern hemisphere of Vesta, as imaged by the Dawn *spaceprobe. This image is dominated by the results of two large impacts: Venenia (the circular feature, slightly off-centre, towards the top) and Rheasilvia (the larger outer impact crater).*

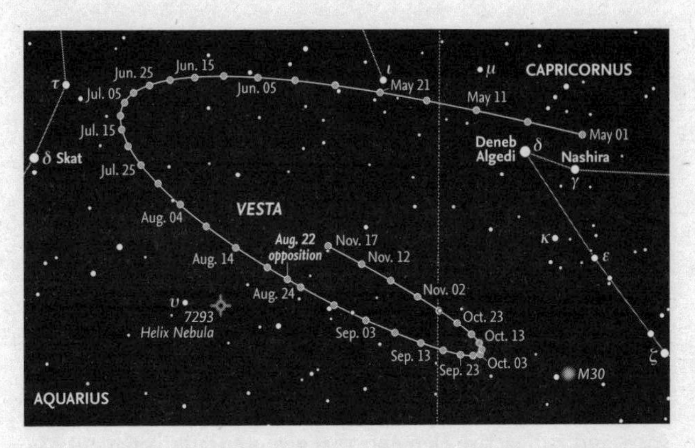

The path of minor planet (4) Vesta around its opposition on August 22 (mag. 5.8). All stars down to magnitude 7.5 are shown.

Meteorite from Vesta

On 2 June 2018, a meteorite fell in Botswana and the fireball it created was observed during its descent. By chance, the parent body (identified as 2018 LA) had been tracked before it entered Earth's atmosphere. This was only the second meteorite that had been detected before the object from which it derived encountered Earth. (The first was a meteorite that fell over the Nubian Desert in Sudan on 7 October 2008. Some 600 fragments of that meteorite were recovered.) The Botswana meteorite (now known as the Motopi Pan meteorite) fell in the Central Kalahari Game Reserve, and 23 fragments

A composite image of the minor planet (4) Vesta as observed by the Dawn *spaceprobe.*

A

were recovered, following an extensive search. Analysis showed that the meteoritic fragments were of the type known as the Howardite-Eucrite-Diogenite meteorites, although with considerable diversity of type. The meteorite is believed to have originated in the large minor planet (4) Vesta through large impacts that occurred in the early history of the Solar System. The fragments seem to have been heated by a very ancient impact (known as the Veneneia impact, which occurred about 4234 million years ago). They persisted on the surface for a very long time, although displaced by another large impact (the Rheasilvia event). Eventually the body was flung into space by another, relatively recent, impact, thought to be the Rubria event dated at a mere 19 (+/- 3) million years. Recent exposure to space is confirmed by investigation of isotopes of noble gases and radioactive elements that give a duration of about 23 million years for the duration of the time the body has been in space. Studies of the orbit determined from entry observations are fully consistent with an origin at Vesta's orbit. Vesta was the minor planet that the *Dawn* spaceprobe visited in July 2011 and then orbited to conduct a 14-month survey.

In 1222, ***Comet Halley*** (1P/Halley) appeared in August. Its path took it through Boötes, Virgo, Libra, and Scorpius. The last recorded observation was on October 23.

September

September – Introduction

The (northern) autumnal equinox occurs on September 23, when the Sun moves south of the equator in the western side of the constellation of Virgo.

Meteors

After the major Perseid shower in August, there is very little shower activity in September. One minor northern shower, known as the ***α-Aurigids***, tends to have two peaks of activity. The principal peak occurs on September 1. In 2022, the Moon is a waxing crescent on that date (First Quarter is on September 3), so conditions are not very favourable. At maximum, the hourly rate reaches 5 to 6 meteors per hour, although the meteors are bright and relatively easy to photograph. Activity from this shower may even extend into October. The ***Southern Taurid*** shower begins this month (on September 10) and, although rates are similarly low, it often produces very bright fireballs. This is a very long shower, lasting until about November 20, with a maximum in 2022 on October 10–11. Unfortunately, this year, activity begins and peaks around Full Moon. As a slight compensation for the lack of shower activity, however, the number of sporadic meteors visible in September reaches its highest rate at any time during the year. Two planets and one minor planet come to opposition during the month.

The ***OSIRIS-REx*** spaceprobe is expected to return its samples from minor planet ***(101955) Bennu*** on 24 September 2023.

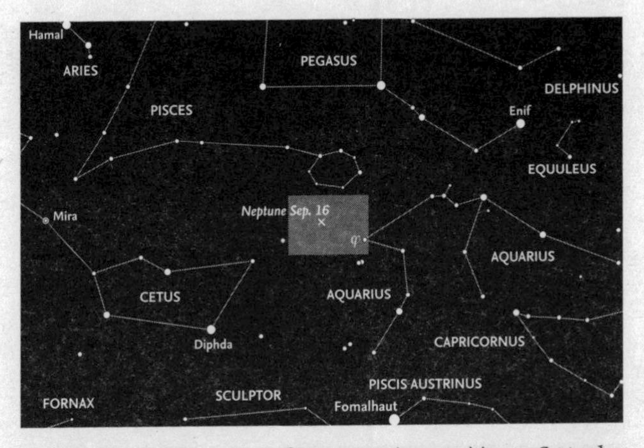

A finder chart for the position of Neptune at its opposition on September 16. The grey area is shown in more detail below.

The path of Neptune in 2022. Stars down to magnitude 7.5 are shown.

S

The planets

Mercury is too close to the Sun to be visible. It actually comes to inferior conjunction on September 23. ***Venus*** is also close to the Sun, but may be seen in the morning twilight early in the month. ***Mars*** is moving eastwards in ***Taurus***, brightening from mag. -0.1 to mag. -0.6 over the month. ***Jupiter*** (mag. -2.9) comes to opposition on September 26 (see the chart below). ***Saturn*** remains in ***Capricornus*** at mag. 0.4 to 0.5. ***Uranus*** is in ***Aries*** at mag. 5.8, and ***Neptune*** in ***Aquarius*** at mag. 7.8, coming to opposition on September 16. Charts showing its position are on page 177. On September 7, minor planet ***(3) Juno*** is at opposition at mag. 7.9 (see the charts on the facing page).

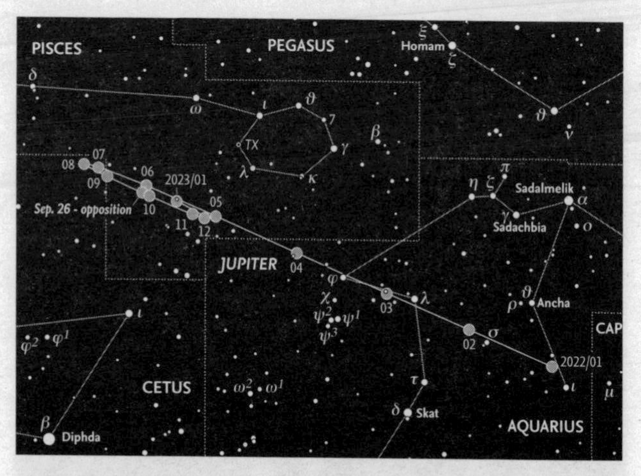

The path of Jupiter in 2022. Stars down to magnitude 6.5 are shown.

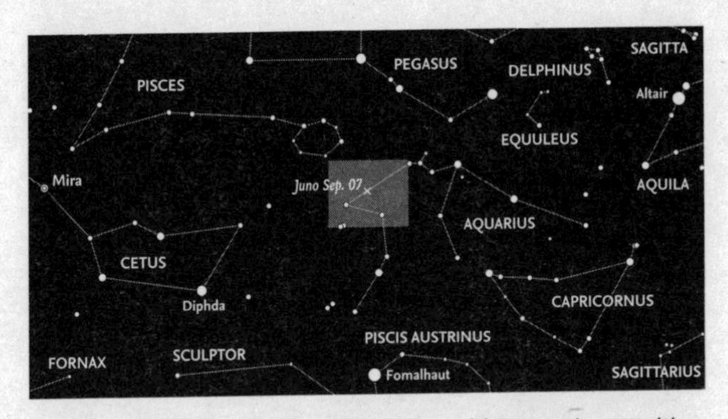

A finder chart for the position of minor planet (3) Juno at its opposition on September 7. The grey area is shown in more detail below.

The path of minor planet (3) Juno around its opposition on September 7, at magnitude 7.9. Background stars are shown down to magnitude 9.0.

S

Sunrise and sunset

City	Date	Sunrise	Sunset
Buenos Aires, Argentina			
	Sep. 01	10:12	21:35
	Sep. 30	09:32	21:56
Cape Town, South Africa			
	Sep. 01	05:05	16:28
	Sep. 30	04:25	16:48
London, UK			
	Sep. 01	05:13	18:48
	Sep. 30	06:00	17:41
Los Angeles, USA			
	Sep. 01	13:27	02:20
	Sep. 30	13:47	01:40
Nairobi, Kenya			
	Sep. 01	03:30	15:35
	Sep. 30	03:19	15:26
Sydney, Australia			
	Sep. 01	20:13	07:37
	Sep. 30	19:33	07:57
Tokyo, Japan			
	Sep. 01	20:13	09:09
	Sep. 30	20:35	08:27
Washington, DC, USA			
	Sep. 01	10:37	23:39
	Sep. 30	11:03	22:52
Wellington, New Zealand			
	Sep. 01	18:45	05:56
	Sep. 30	17:56	06:25

NB: the times given are in Universal Time (UT)

The Moon's phases and ages

Northern hemisphere

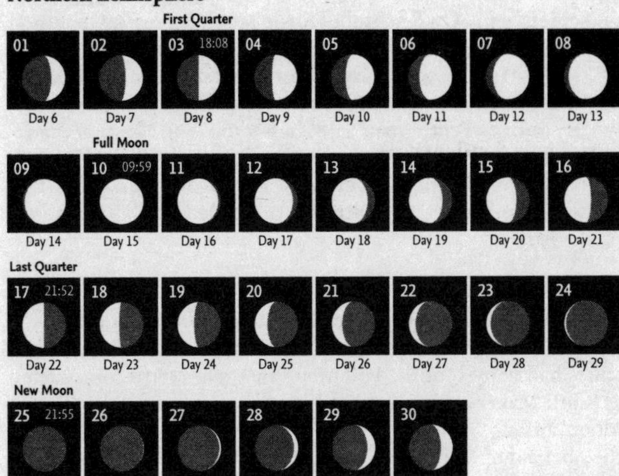

Southern hemisphere

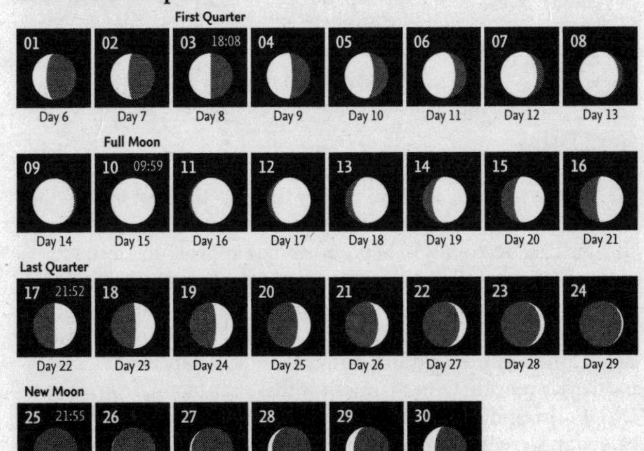

S

The Moon

The First Quarter Moon is 2.5° north of ***Antares*** in ***Scorpius*** on September 3. On September 8, the Moon passes 3.9° south of ***Saturn***. On September 11, one day past Full, the Moon is 1.8° south of ***Jupiter***. Just before Last Quarter, the Moon is 3.6° north of ***Mars*** on September 17. On September 20, the Moon is 1.9° south of ***Pollux*** (mag. 1.18) in ***Gemini***. The Moon is close to ***Venus*** on September 25, low in the morning twilight. On September 30, the Moon again passes north of ***Antares*** as it did at the beginning of the month, but this time very slightly closer (2.3° north).

Corn Moon

To the peoples of North America, September was particularly important because corn (maize) was ready for harvest. Many tribes had names for the Full Moon that referred to corn, such as 'Corn Maker Moon' among the Abenaki of northern Maine; 'Middle Moon between harvest and eating corn' to the Algonquin in the Northeast and Great Lakes area, although it was the 'Moon when freeze begins on stream's edge' for the Cheyenne of the Great Plains.

On 5 September 1977, the ***Voyager 1*** spaceprobe was launched. It studied Jupiter, Saturn, and Saturn's satellite, Titan. It became the first object to leave the heliosphere and enter interstellar space.

In Europe, the September Full Moon was generally called the 'Harvest Moon', and technically this was the first Full Moon after the autumnal equinox (23 September in 2022). In general, this Full Moon comes in September, but approximately once every three years, Full Moon comes in October, when it is that Full Moon that is known as the 'Harvest Moon'. The term 'Harvest Moon' was the only name for the Full Moon that was determined by the equinox, rather than being specific to any particular month. Other names for this Full Moon, from the Old World, and specifically from the Old English/Anglo-Saxon calendar, were 'Full Corn Moon' – in this case 'corn' meaning wheat or barley – as well as 'Barley Moon'.

The *Pioneer 10* spaceprobe

The two *Pioneer* spaceprobes were the first specifically designed to investigate the outer Solar System. They were also the first objects to achieve the necessary velocity to escape from the Sun's influence. (The two *Voyager* and the *New Horizons* spaceprobes have also attained the necessary velocity.) Both *Pioneer* spaceprobes carried the famous plaque, designed to indicate to any potential aliens, who might encounter the probes, the origin of the probes. *Pioneer 10* was launched on 2 March 1972. It was the first spaceprobe to encounter Jupiter, and began photography on 6 November 1973. Closest approach was on 3 December 1973.

One of the best images of Jupiter obtained by Pioneer 10. *Although primitive when compared with later images, it was one of the first close-up images of the giant planet.*

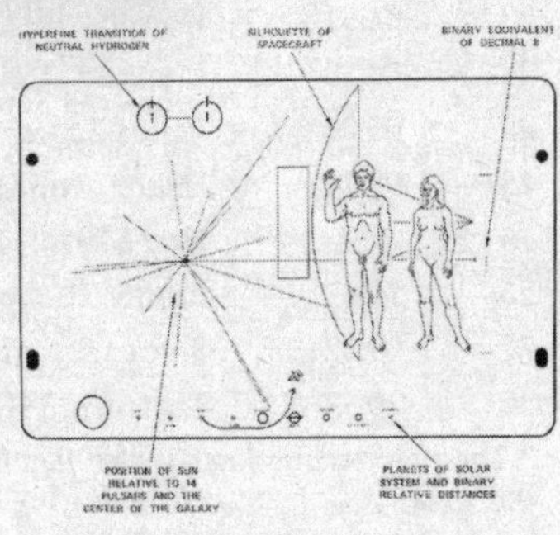

The plaque carried by both Pioneer *spaceprobes, with added annotation to indicate (to humans) how the diagram should be interpreted.*

Calendar for September

01		α-Aurigid shower maximum
03	15:29	Antares 2.5°S of Moon
03	18:08	First Quarter
05	01:00 *	Venus is 0.8°N of Regulus
07	16:49	Minor planet (3) Juno at opposition (mag. 7.9)
07	18:19	Moon at perigee = 364,492 km
08	10:31	Saturn 3.9°N of Moon
09	01:00 *	Mars 4.3°N of Aldebaran
10–Nov.20		Southern Taurid meteor shower
10	09:59	Full Moon
10	18:54	Neptune 3.0°N of Moon
11	15:17	Jupiter 1.8°N of Moon
14	22:59	Uranus 0.8°S of Moon
16	18:19	Aldebaran 7.8°S of Moon
16	22:21	Neptune at opposition (mag. 7.8)
17	01:43	Mars 3.6°S of Moon
17	21:52	Last Quarter
19	14:43	Moon at apogee = 404,556 km
20	08:17	Pollux 1.9°N of Moon
23	01:04	September equinox
23	04:51	Regulus 4.8°S of Moon
23	06:50	Mercury at inferior conjunction
25	05:08	Venus 2.8°S of Moon
25	08:14	Mercury 6.7°S of Moon
25	21:55	New Moon
26	19:33	Jupiter at oposition (mag. -2.9)
27	09:57	Spica 4.2°S of Moon
30	20:53	Antares 2.3°S of Moon

* *These objects are close together for an extended period around this time.*

September 12 • *The Moon and Jupiter are close together, high in the northern sky (as seen from Sydney).*

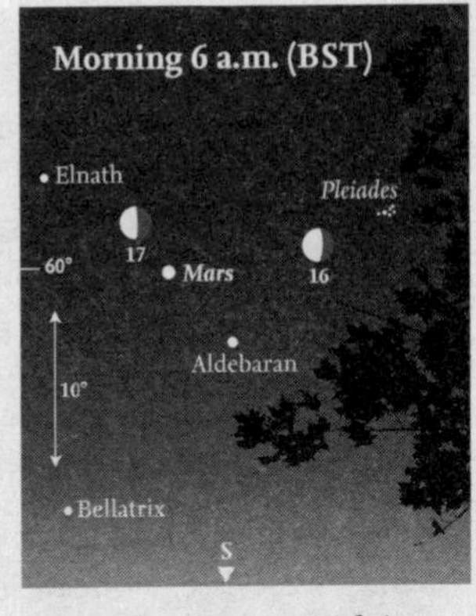

September 16–17 • *High in the south, the Moon passes the Pleiades, Aldebaran, Mars and Elnath (as seen from London).*

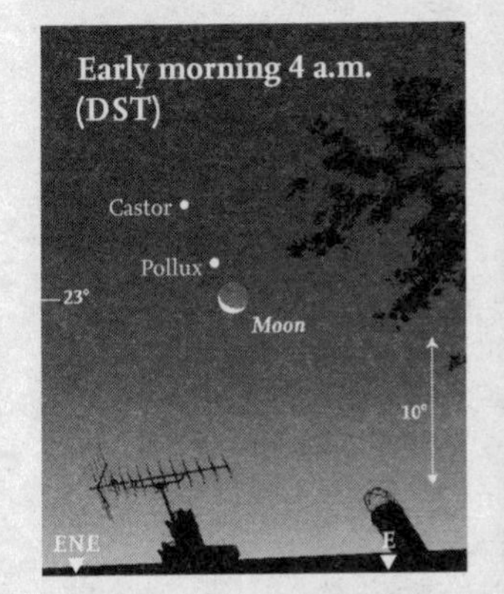

September 20 • *The Moon lines up with Castor and Pollux (as seen from central USA).*

September 30 – October 1 • *The waxing crescent Moon passes between Antares and Sabik (η Oph). As seen from Sydney.*

S

Early September 23:00 — Mid September 22:00 — Late September 21:00

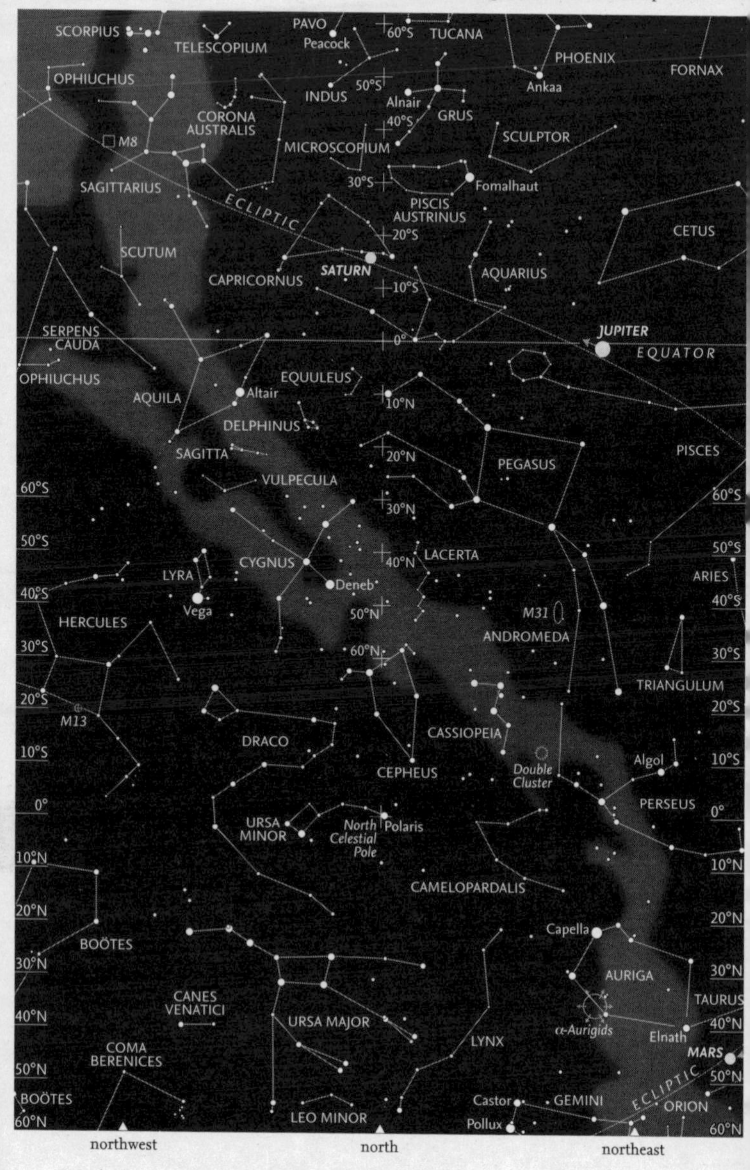

September – Looking North

For mid-northern observers, ***Cassiopeia*** is high overhead, and ***Cepheus*** is 'upside-down', apparently hanging from the Milky Way, high above the Pole. ***Perseus*** is high in the northeast, at about the same altitude as ***Ursa Minor*** and ***Polaris***.

For those same observers, ***Ursa Major*** is now 'right way up', low in the south, although some of the southernmost stars are lost along or below the horizon. ***Auriga***, with bright ***Capella***, and the small triangle of the 'Kids' is clearly visible in the northeast. Higher in the sky, above Perseus, the whole of ***Andromeda*** is visible, together with the small constellation of ***Triangulum*** and, still higher, the Great Square of ***Pegasus***, and beyond it, most of the zodiacal constellation of ***Pisces***.

The clouds of stars forming the Milky Way are not particularly striking in Auriga and Perseus, but beyond Cassiopeia, and on towards ***Cygnus***, they become much denser and easier to see.

Cygnus is high in the northwest and ***Deneb***, its principal star, is close to the zenith for observers at 40–50° north, with ***Lyra*** and bright ***Vega***, slightly farther west. Farther towards the south, most of ***Hercules*** is clearly visible, with the 'Keystone' and the globular cluster M13. The head of ***Draco*** lies between Hercules and Cepheus and the whole of that constellation is easily seen as it curls around Ursa Minor and the North Celestial Pole.

Observers in the far north may be able to detect ***Castor*** (α Geminorum) skimming the northern horizon, although brighter ***Pollux*** (β Geminorum) will be too low to be detected until later in the night.

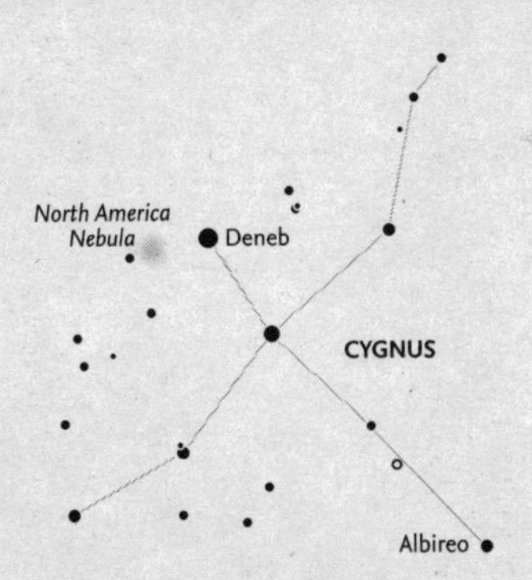

The constellation of Cygnus. Deneb (α Cyg) is one of the stars forming the Summer Triangle. The other two are Vega (α Lyr) and Altair (α Aql).

CEPHEUS
60°N
HERCULES
TRIANGULUM
M31
50°N
Vega
Deneb
ANDROMEDA
40°N
CYGNUS
ARIES
LYRA
LACERTA
30°N
VULPECULA
PEGASUS
20°N
SAGITTA
PISCES
DELPHINUS
10°N
Altair
ECLIPTIC
EQUULEUS
OPHIUCHUS
0°
EQUATOR
JUPITER
AQUILA
10°S
AQUARIUS
SATURN
SCUTUM
20°S
CAPRICORNUS
CETUS
SERPENS CAUDA
60°N
Fomalhaut
30°S
60°N
50°N
PISCIS AUSTRINUS
MICROSCOPIUM
SCULPTOR
SAGITTARIUS
M8
50°N
40°S
40°N
Alnair
40°N
Ankaa
GRUS
50°S
INDUS
CORONA AUSTRALIS
PHOENIX
OPHIUCHUS
30°N
60°S
Peacock
30°N
FORNAX
TELESCOPIUM
SCORPIUS
20°N
TUCANA
ARA
20°N
Achernar
PAVO
10°N
SMC
10°N
ERIDANUS
HYDRUS
OCTANS
NORMA
0°
HOROLOGIUM
South Celestial Pole
TRIANGULUM AUSTRALE
0°
APUS
RETICULUM
MENSA
10°S
CHAMAELEON
CIRCINUS
Rigil Kentaurus
LUPUS
10°S
LMC
DORADO
Tarantula N.
Hadar
20°S
MUSCA
Acrux
Mimosa
20°S
CAELUM
PICTOR
Canopus
VOLANS
30°S
CARINA
CRUX
Omega Centauri
30°S
Eta Car Nebula
40°S
CENTAURUS
COLUMBA
40°S
VELA
PUPPIS
50°S
HYDRA
50°S
Adhara
PYXIS
LEPUS
CANIS MAJOR
ANTLIA
CORVUS
60°S
60°S
southeast
south
southwest

September – Looking South

The three stars forming the apices of the (northern) Summer Triangle, ***Deneb*** (α Cygni), ***Vega*** (α Lyrae), and ***Altair*** (α Aquilae) are prominent in the southwest. The Great Rift is clearly visible, starting near Deneb and running down the centre of the Milky Way towards the centre of the Galaxy in ***Sagittarius***, and beyond into ***Scorpius***, only petering out in ***Centaurus*** and ***Crux***.

Most of the constellation of ***Capricornus*** lies just west of the meridian, with ***Aquarius*** just slightly farther north on the eastern side. The next zodiacal constellation, ***Pisces***, is clearly seen, including the prominent asterism, known as the 'Circlet'. Much of the constellation of ***Cetus*** is visible south of it. South of Aquarius is ***Piscis Austrinus***, with its single bright star, ***Fomalhaut***, in an otherwise fairly barren area of sky. Still slightly farther south is the undistinguished constellation of ***Sculptor***, and, closer to the meridian, the line of stars that forms part of ***Grus***.

For observers south of the equator, Sagittarius is high in the west. Below it is the curl of stars that is ***Corona Australis***. Below (south of) Sagittarius is the tail and 'sting' of Scorpius. Still farther south is the constellation of ***Lupus*** and the scattered stars of ***Centaurus***, with ***Rigil Kentaurus*** and ***Hadar*** (α and β Centauri, respectively). Between the 'sting' of Scorpius and those two bright stars are the small constellations of ***Ara*** and ***Triangulum Australe***. Next comes Crux itself. If the long axis of the cross is extended right across the sky over the largely empty area of sky around the South Celestial Pole, it points in the general direction of ***Achernar*** (α Eridani). Between Achernar and Ara, west of the meridian, lies the constellation of ***Pavo*** with it sole bright star ***Peacock*** (α Pavonis). To the south, the brightest star of the southern hemisphere, ***Canopus*** (α Carinae) is hugging the horizon for observers at the latitude of Sydney in Australia.

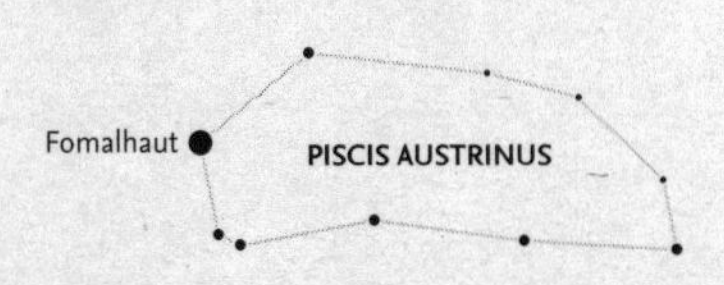

For observers at 30°S, the constellation of Piscis Austrinus, with its single bright star, Fomalhaut, is almost overhead.

S

On 5 September 1977, the ***Voyager 1*** spaceprobe was launched. It studied Jupiter, Saturn, and Saturn's satellite, Titan. It became the first object to leave the heliosphere and enter interstellar space.

October

October – Introduction

There is one major meteor shower and three minor ones that are active during October.

The ***Orionids*** are the major, fairly reliable, meteor shower that may be observed in October. Like the May ***η-Aquariid*** shower, the ***Orionids*** are associated with Comet 1P/Halley. During this second pass through the stream of particles from the comet, slightly fewer meteors are seen than in May, but conditions are more favourable for northern observers. In both showers the meteors are very fast, and many leave persistent trains. Although the Orionid maximum is quoted as October 21–22 for 2022, in fact there is a very broad maximum, lasting about a week roughly centred on that date, with hourly rates around 25. Occasionally, rates are higher (50–70 per hour). In 2022, the Moon is a waning crescent at the nominal maximum, so conditions are reasonably favourable.

The *Cassini-Huygens* spaceprobe mission was launched on 15 October 1997.

The faint shower of the ***Southern Taurids*** (often with bright fireballs) peaks on October 10–11. The Southern Taurid maximum occurs one day after Full Moon, so conditions are particularly unfavourable, compared with the Orionids. Towards the end of the month (around October 20), another shower (the ***Northern Taurids***) begins to show activity, which peaks early in November. The parent comet for both Taurid showers is Comet 2P/Encke, which has the shortest period (3.3 years) of any major comet. The meteors in both Taurid streams are relatively slow and bright.

A short, minor northern shower, the ***Draconids***, begins on October 6 and peaks on October 8–9. Although the rate is about 10 meteors per hour, the whole shower occurs near Full Moon in 2022, so observing conditions are extremely poor.

The ***OSIRIS-REx*** spaceprobe successfully collected a sample from minor planet ***(101955) Bennu*** on 20 October 2020.

The planets

Mercury is at greatest elongation west at mag. -0.6 on October 8, low and difficult to observe in the morning sky. ***Venus*** is close to the Sun, and comes to superior conjunction on October 22. It is occulted by the Moon on October 25, but the event occurs in daylight over Africa and the Atlantic, so is basically unobservable. ***Mars***, in ***Taurus***, brightens from mag. -0.6 to mag. -1.2 over the month. It begins retrograde motion on October 31. ***Jupiter***, after opposition in September, is still retrograding in ***Pisces*** at mag. -2.9 to -2.8. ***Saturn*** is retrograding slowly in ***Capricornus*** and ***resumes*** direct eastwards motion on October 23. ***Uranus*** (mag. 5.7) is in ***Aries***, and ***Neptune*** (mag. 7.8) remains in ***Aquarius***.

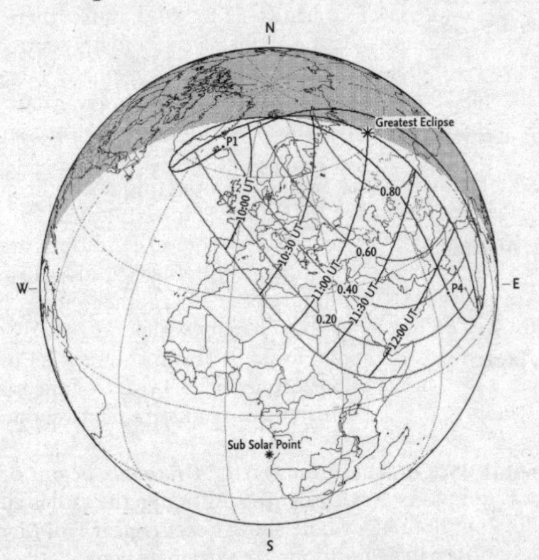

The partial solar eclipse of 25 October 2022. The greatest eclipse occurs over northern Siberia, but the partial phases will be visible over a wide area of Asia and (in particular) eastern Europe. Western Europe and East Africa will be able to see 20 to 40 per cent of the Sun's disc hidden by the Moon.

O

Sunrise and sunset

City	Date	Sunrise	Sunset
Buenos Aires, Argentina			
	Oct. 01	09:30	21:57
	Oct. 31	08:53	22:22
Cape Town, South Africa			
	Oct. 01	04:23	16:49
	Oct. 31	03:47	17:13
London, UK			
	Oct. 01	06:02	17:39
	Oct. 31	06:53	16:36
Los Angeles, USA			
	Oct. 01	13:48	01:39
	Oct. 31	14:12	01:02
Nairobi, Kenya			
	Oct. 01	03:19	15:26
	Oct. 31	03:12	15:21
Sydney, Australia			
	Oct. 01	19:32	07:58
	Oct. 31	18:55	08:22
Tokyo, Japan			
	Oct. 01	20:36	08:25
	Oct. 31	21:02	07:47
Washington, DC, USA			
	Oct. 01	11:04	22:51
	Oct. 31	11:34	22:09
Wellington, New Zealand			
	Oct. 01	17:54	06:26
	Oct. 31	17:08	07:00

NB: the times given are in Universal Time (UT)

The Moon's phases and ages

Northern hemisphere

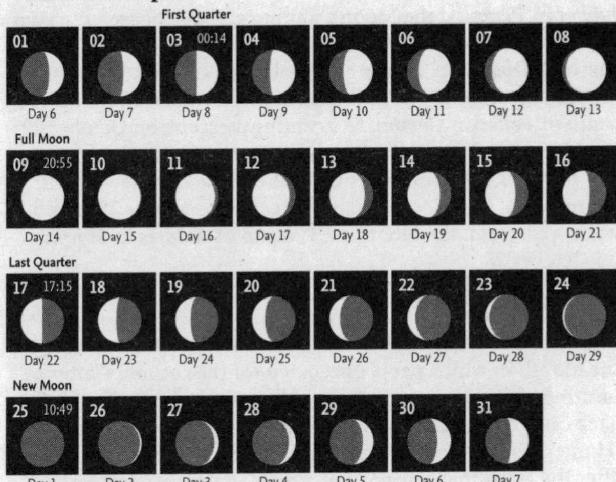

Southern hemisphere

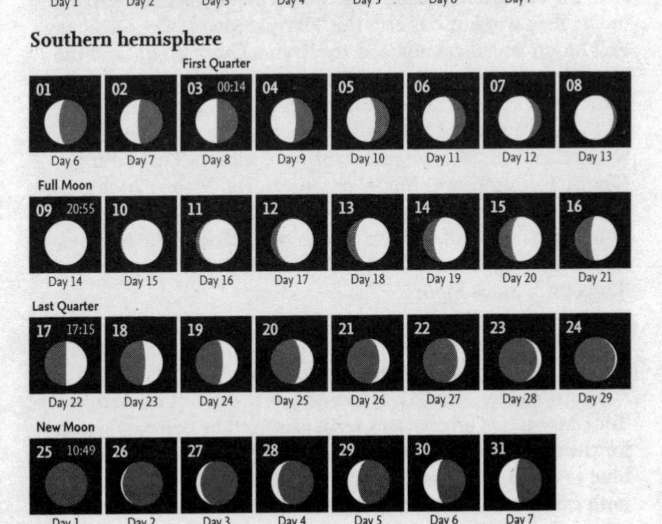

O

The Moon

On October 8, one day before Full Moon, it is 2.1° south of ***Jupiter*** in ***Pisces***. At the Moon's waning gibbous phases, it passes 0.9° north of ***Uranus*** (mag. 5.7) in ***Aries*** on October 12, then passes ***Aldebaran*** in ***Taurus*** on October 14, and is 3.6° north of ***Mars*** the next day. At Last Quarter, on October 17, it is 1.8° south of ***Pollux*** in ***Gemini***. As a waning crescent, on October 20 the Moon is 4.9° north of ***Regulus***, between it and ***Algieba***. It is 4.2° north of ***Spica*** in ***Virgo*** on October 24, and at New Moon, the next day, there is a partial solar eclipse (see page 193). As a waxing crescent, the Moon is 2.3° north of ***Antares*** in ***Scorpius*** on October 28.

Hunter's Moon

In the northern hemisphere of the Old World, October was the month in which people prepared for the coming winter by hunting, slaughtering livestock and preserving meat for food. This caused the Full Moon in October to become known as the 'Hunter's Moon'. Every three years, however, the first Full Moon after the autumnal equinox fell, not in September, but early in October, when it was also the 'Harvest Moon'. The October Full Moon was also known as the 'Dying Grass Moon' and the 'Blood' or 'Sanguine Moon'. ('Blood Moon' is also the term sometimes applied to the Moon during a lunar eclipse.)

In the New World there was a great variety of names. Many were related to the changes in autumn, such as 'Leaf-falling Moon', 'Falling Leaves Moon' or simply 'Fall Moon'. To the Algonquin of the Northeast and Great Lakes area it was the 'White Frost on Grass Moon'. The Assiniboine of the Northern Plains had a rather different sort of name. To them it was the 'Joins Both Sides Moon'.

The colour of the Moon

As with the term 'supermoon', another North American term has been widely adopted. This involves the use of the term 'Blue Moon'. Originally, this term was used by meteorologists for the optical phenomenon where the Moon literally appears blue in colour. The same effect may occur with the Sun, and in both cases was (and still is) caused by particles suspended in the upper atmosphere that scatter the appropriate wavelengths (colours) of sunlight. (This rare event is the actual origin of the phrase 'Once in a blue moon'.) The most common source of such particles in the upper atmosphere, which need to be of a specific size, is a wildfire. On occasions, forest fires in Canada have produced so many particles and spread them so widely downwind that tinted suns and moons have been seen even in Europe. A somewhat similar effect may sometimes occur when volcanoes eject large quantities of ash into the upper atmosphere, although the colours tend to be orange and red, because of a different range of particle sizes. Orange and red colours are often associated, in Europe, with incursions of particles from the North African deserts. A different range of particle sizes from forest fires (in particular) may give rise to a green coloration of the Moon or Sun, but this occurrence is exceptionally rare.

Increasingly, however, the term 'Blue Moon' came to be applied to the third Full Moon of the four that occur in a meteorological season or to the second Full Moon that occurred in a particular calendar month. Nowadays, the latter usage is frequently applied by the media to the second Full Moon in any month. It is this usage that has crossed the Atlantic to be taken up and publicized by European media.

In reality, of course, the colour of the Moon is very subdued. To the eye, overall, it is simply different shades of grey. There are only lighter and darker areas (the highlands and the maria, respectively). With optical aid, a very skilled observer can detect differences in colour in a few areas, such as a slight reddish tint to part of the floor of the crater Fracastorius, caused by the mineral variations on the Moon's surface.

The colour of the Moon, as seen by the naked eye, is grey – various shades of grey.

Calendar for October

02–Nov.07		Orionid meteor shower
03	00:14	First Quarter
04	16:34	Moon at perigee = 369,325 km
06–10		Draconid meteor shower
08–09		Draconid meteor shower maximum
08	02:33	Neptune 3.1°N of Moon
08	18:12	Jupiter 2.1°N of Moon
08	21:14	Mercury at greatest elongation (18.0°W, mag. -0.6)
09	20:55	Full Moon
10–11		Southern Taurid meteor shower maximum
12	06:46	Uranus 0.9°S of Moon
14	02:56	Aldebaran 8.0°S of Moon
15	04:31	Mars 3.6°S of Moon
17	10:20	Moon at apogee = 404,328 km
17	16:19	Pollux 1.8°N of Moon
17	17:15	Last Quarter
20–Dec.10		Northern Taurid meteor shower
20	13:17	Regulus 4.9°S of Moon
21–22		Orionid meteor shower maximum
22	21:17	Venus at superior conjunction
24	15:44	Mercury 0.4°S of Moon
24	18:14	Spica 4.2°S of Moon
25	10:49	New Moon
25	11:00	Partial solar eclipse
25	12:05	Venus occultation
28	03:21	Antares 2.3°S of Moon
29	14:36	Moon at perigee = 368,291 km

October 8 • *The almost Full Moon with Jupiter, low in the eastern sky (as seen from central USA).*

October 12–14 • *The Moon passes the Pleiades, Aldebaran, Mars and Elnath. (as seen from central USA).*

October 18 • *The Moon in the company of Castor and Pollux, in the northeastern sky (as seen from Sydney).*

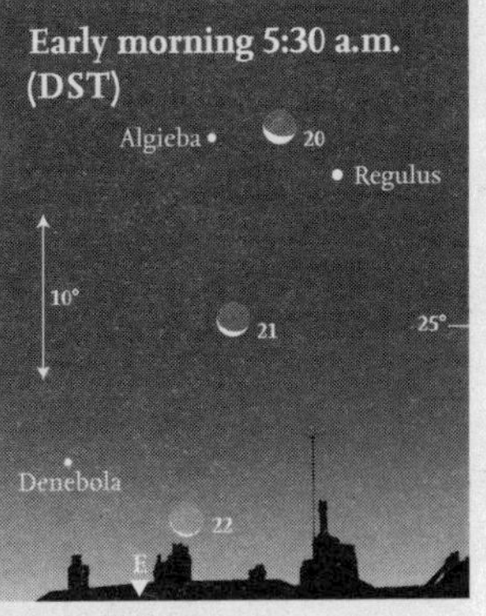

October 20–22 • *The Moon passes through the constellation of Leo, the Lion (as seen from central USA).*

O

Early October 23:00 — Mid October 22:00 — Late October 21:00

October – Looking North

The Milky Way arches across the northern sky, running from ***Auriga*** in the east to ***Cygnus*** and ***Aquila*** in the west. The constellations of ***Perseus***, ***Cassiopeia***, and ***Cepheus*** lie along it. For observers in middle northern latitudes, these constellations are high overhead. (To observers in the far north, Cassiopeia is near the zenith.) ***Andromeda*** is high in the north, beyond Perseus and the other side of the Milky Way. The two small constellations of ***Triangulum*** and ***Aries*** are to the south of it. In the northwest, the constellation of ***Lyra*** lies farther south, clear of the Milky Way.

Between Cassiopeia and Cygnus lies the zig-zag of faint stars that form the small constellation of ***Lacerta***, which is often difficult to recognize because it lies across the Milky Way. Two of the stars forming the Summer Triangle, ***Deneb*** (α Cygni) and ***Vega*** (α Lyrae) are readily visible, but the third, ***Altair*** (α Aquilae), is approaching the northwestern horizon. So too, much farther north, is the constellation of ***Hercules***. The head of ***Draco*** is at about the same altitude as ***Polaris*** in ***Ursa Minor***, as is ***Capella*** (α Aurigae). The large constellation of ***Ursa Major*** is now directly beneath the Pole, visible above the horizon to the north. ***Castor*** and ***Pollux*** (α and β Geminorum, respectively) are in the northeast, while the northernmost stars of ***Boötes*** and the circlet of ***Corona Borealis*** are in the northwest. The brightest star in Boötes, orange-tinted Arcturus, is below the horizon early in the night.

The long chain of faint stars that is the constellation of ***Lynx***, runs almost vertically between Ursa Major and ***Gemini***, with the other faint constellation of ***Camelopardalis*** between it and Perseus.

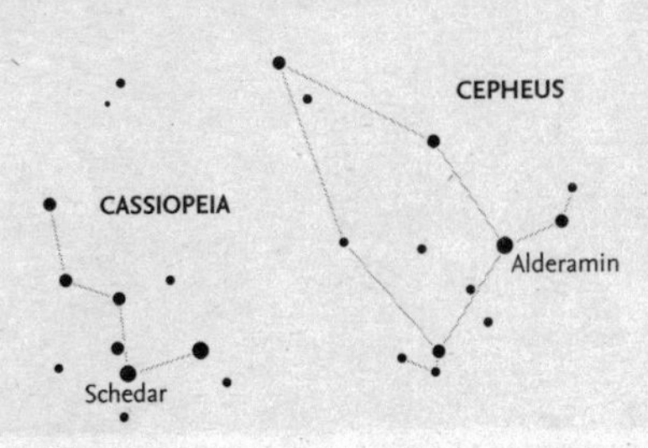

This month two of the northern circumpolar constellations, Cassiopeia and Cepheus, are almost at the meridian.

O

Early October 23:00 — Mid October 22:00 — Late October 21:00

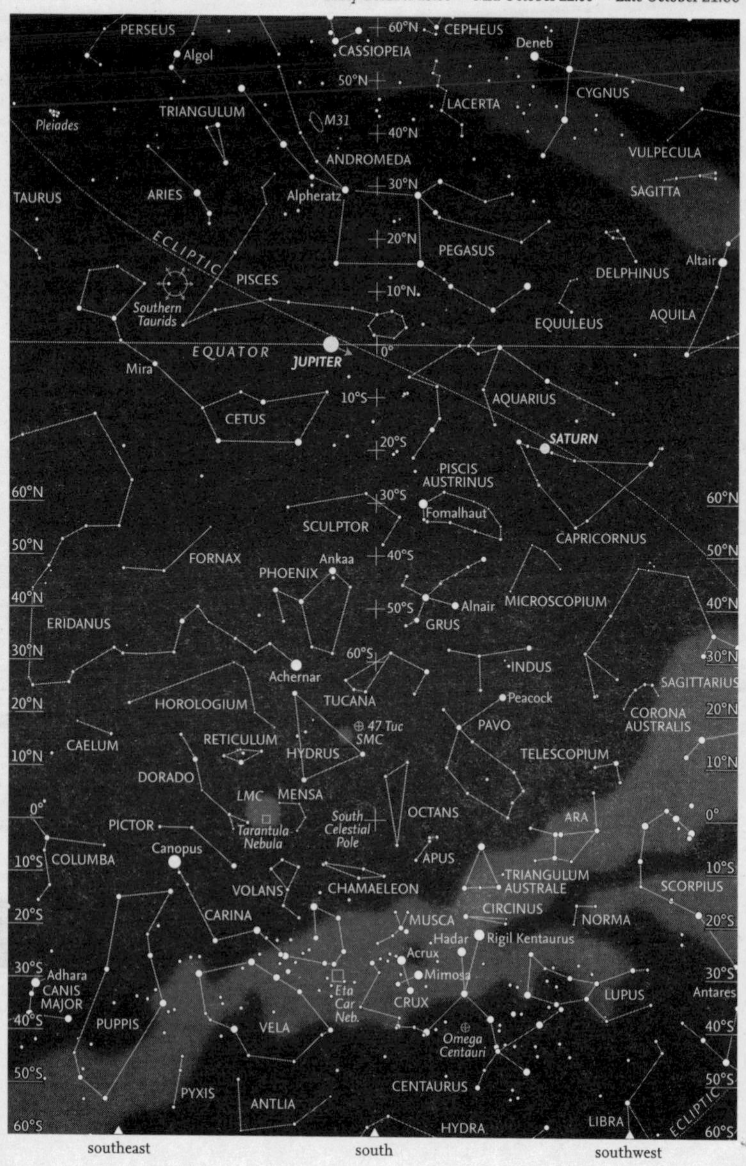

October – Looking South

The Great Square of ***Pegasus*** is now on the meridian due north, with the constellation of ***Andromeda*** stretching away from it in the northeast. The star at the northeastern corner of the Great Square is actually ***Alpheratz*** (α Andromedae). The two parts of the zodiacal constellation of ***Pisces*** are to the south and east of Pegasus. ***Aquarius*** and the non-zodiacal constellation of ***Cetus*** are slightly farther south while ***Capricornus*** is sinking in the west.

South of Aquarius is ***Pisces Austrinus*** with its single bright star, ***Fomalhaut***, while straddling the meridian slightly east of it is the faint constellation of ***Sculptor***. Below (south) these two constellations are the roughly cross-shaped constellation of ***Grus*** with, on the other side of the meridian, ***Phoenix***. Just below the constellation of Pheonix is ***Achernar*** (α Eridani), the bright star that ends the long winding constellation of ***Eridanus*** that actually begins near Rigel in Orion. Below Achernar is the triangular constellation of ***Hydrus***, with the ***Small Magellanic Cloud*** (SMC) on its northwestern side. At about the same altitude towards the west is the constellation of ***Pavo*** with ***Peacock*** (α Pavonis).

The ***Large Magellanic Cloud*** (LMC) lies southeast of Hydrus, partly in each of the two faint constellations of ***Dorado*** and ***Mensa***. The Milky Way lies in the east, where portions of ***Sagittarius*** and ***Scorpius*** are clearly visible. Roughly level with the 'Sting' of Scorpius are the constellations of ***Ara*** and ***Triangulum Australe***. Farther east are the undistinguished constellations of ***Apus***, ***Chamaeleon*** and ***Volans***. Even farther to the east is brilliant ***Canopus*** (α Carinae), the second brightest star in the sky (after Sirius) at mag. -0.6.

For observers at about 30° south (roughly the latitude of Sydney in Australia), the two brightest stars of ***Centaurus*** (***Rigil Kentaurus*** and ***Hadar***) are brushing the horizon, while the small constellation of ***Crux*** is lost below it. Only observers even farther south will be able to see that constellation in full, together with the constellation of ***Lupus***, the full extent of Centaurus and the large constellation of ***Vela***, with ***Puppis*** in the southeast.

Galileo and Cassini-Huygens

The Galileo spaceprobe

The ***Galileo*** project was designed to study Jupiter. It consisted of an orbiter and an atmospheric probe. It was launched on the Space Shuttle *Atlantis* on 18 October 1989. After two gravitational 'slingshot' manoeuvres, one of Venus and the other of Earth to alter its trajectory and increase its velocity, it passed through the asteroid belt between Mars and Jupiter. It obtained the first close-up image of a minor planet ***(951) Gaspra***, see page 162) and discovered the first satellite of a minor planet, finding ***Dactyl*** orbiting ***(243) Ida*** (see page 162). The probe was ideally placed to observe the collision of ***Comet Shoemaker-Levy 9*** with Jupiter between 16 and 22 July 1994. The entry probe was separated from the orbiter on 13 July 1995, and descended into the Jovian atmosphere on 7 December 1995, returning extremely useful scientific observations. This next day, 8 December 1995, the orbiter was successfully placed in orbit around the planet.

The orbiter was placed in such an orbit that it was able to conduct observation of the satellites of Jupiter as well as of the planet itself. Extensive observation of the four major satellites was carried out, together with observations of smaller satellites and Jupiter's radiation environment. The primary mission ended on 7 December 1997, but an extension was agreed, and this later phase lasted until 31 December 1999. To avoid contamination of the planet – Galileo had not been fully sterilized before launch – the orbiter was deliberately plunged into the atmosphere at a high velocity, to cause it to burn up completely, on 21 September 2003.

The Cassini-Huygens mission

A combined mission was mounted to study Saturn and its major satellite, Titan. *Cassini* itself was contributed by NASA and designed to orbit Saturn and study the planet, its rings and satellites. The *Huygens* probe, a joint European Space Agency and Italian Space Agency project, was a Titan lander.

The mission was launched on 15 October 1997, and had three gravitational 'slingshot' encounters with Venus (twice) and Earth (once). In travelling to Saturn it also flew by the minor planet (2685) Masursky and had an encounter with Jupiter in December 2000. The orbiter entered orbit around Saturn on 1 July 2004, after passing through a gap in the ring system. The *Huygens* probe separated from the *Cassini* mothership on 25 December 2004,

The *Huygens* probe successfully parachuted down onto the surface of Titan on 14 January 2005, returning very significant scientific results. Although spaceprobes have landed on minor planets and a comet, this is the only landing on a planetary satellite, other than on Earth's Moon, and the only one in the outer Solar System. It was particularly important because Titan is the only satellite to have a dense atmosphere; so dense, in fact, that hazes prevent the surface from being visible.

The orbiter continued its studies of Saturn, its ring system and satellites, making numerous very important discoveries, including atmospheric features, new satellites, structures in the rings, and the electromagnetic environment of the planet. The orbiter's trajectory took it several times past Titan, and radar studies revealed many important features of the satellite, including liquid lakes and dune fields.

A flyby of the satellite Enceladus revealed a thin atmosphere and 'geysers', erupting from the south polar region, giving rise to the suggestion that the satellite has a sub-surface ocean or significant pockets of liquid water.

To avoid any potential contamination of the satellites that might harbour alien lifeforms and possible contamination of the planet's atmosphere, a controlled destruction was agreed. The main Cassini orbiter passed between the inner ring and the planet on 26 April 2017. The mission ended on 15 September 2017, when the orbiter plunged into the outer atmosphere.

November

November – Introduction

There are three meteor streams that begin in earlier months, but continue into November. There is an additional major shower that begins in this month, which may sometimes show extraordinary bursts of activity. The ***Southern Taurid*** shower, which began on September 10, may continue until November 20. Similarly, the ***Orionids***, beginning on October 2, continue until November 7. (Both of these showers are described in earlier months.)

The third of these enduring meteor streams, the ***Northern Taurid*** shower, begins in late October (October 20), reaches maximum – although with just a low rate of about five meteors per hour – on November 12–13. Full Moon is on November 8, so conditions are not favourable at the shower's maximum, although quite good for the early part of the shower, around New Moon. The shower gradually trails off, ending around December 10. There is an apparent 7-year periodicity in fireball activity, but 2022 is unlikely to be a peak year.

Far more striking, however, are the ***Leonids***, which have a relatively short period of activity (November 6–30), with maximum on November 17–18. This shower is associated with Comet 55P/Tempel-Tuttle, which has an orbital period of 33.22 years. The shower has shown extraordinary activity on various occasions with many thousands of meteors per hour. High rates were seen in 1999, 2001 and 2002 (reaching about 3000 meteors per hour) but have fallen dramatically since then. The rate in 2022 is likely to be about 15 per hour. Because of the radiant's location, this shower is best seen from the northern hemisphere. These meteors are the fastest shower meteors ever recorded (about 70 km per second) and often leave persistent trains. Apart from the sheer numbers occasionally seen, the shower is very rich in faint meteors. In 2022, maximum is just after Last Quarter, so conditions are not particularly favourable.

There is a somewhat enigmatic, minor southern meteor shower that begins activity in late November (nominally November 28). This is the ***Phoenicids***. Little is known of this shower, partly because the parent comet is believed to be the disintegrated comet 289P/Blanpain. With no accurate knowledge of the location of the remnants of the comet, predicting the possible rate becomes little more than guesswork. The rate is variable and may rapidly increase if the orbit is nearby. There is a tendency for bright meteors to be frequent and all the meteors are fairly slow. The radiant is located within the constellation of ***Phoenix***, not far from the border with ***Eridanus*** and the bright star ***Achernar*** (α Eridani).

The planets

Mercury passes superior conjunction on the far side of the Sun on November 8. ***Venus***, although bright (mag. -4.0 to -3.9) is lost in twilight, too close to the Sun to be readily visible. ***Mars***, which began its westward retrograde motion on October 31, is in ***Taurus***, brightening from mag. -1.2 to -1.8 over the month. ***Jupiter*** (mag. -2.8 to -2.6), initially retrograding in ***Pisces***, resumes direct motion on November 26. ***Saturn*** (mag. 0.7 to 0.8) is moving slowly eastwards (i.e., with direct motion) in ***Capricornus***. ***Uranus***, in ***Aries*** at mag. 5.6 is at opposition on November 9 (see the charts below). ***Neptune*** is in ***Aquarius*** at mag. 7.8 to 7.9.

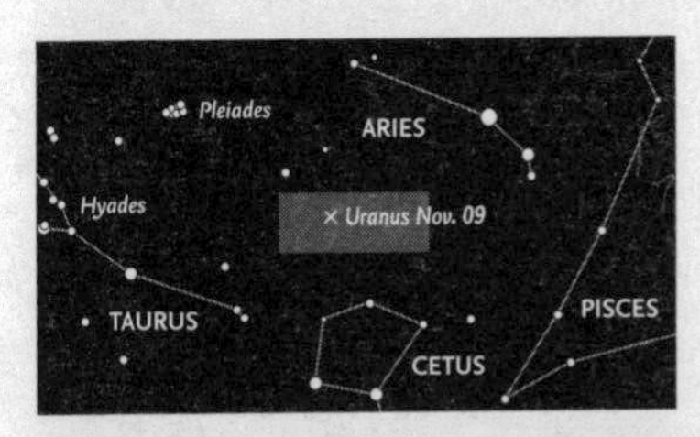

The position of Uranus at its opposition on November 9. The grey area is shown in more detail on the chart below. On that chart, stars down to magnitude 7.5 are shown.

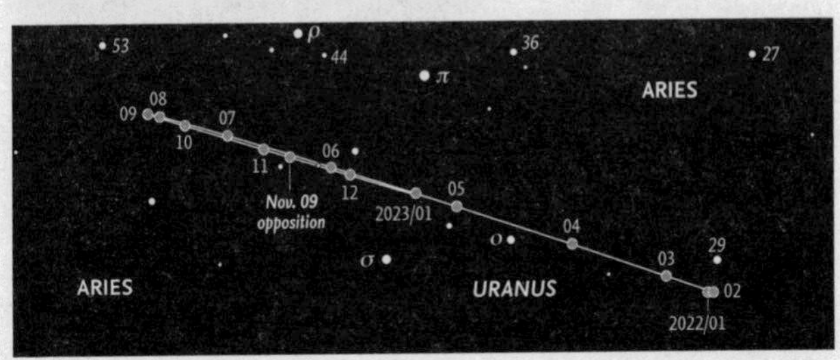

Sunrise and sunset

City	Date	Sunrise	Sunset
Buenos Aires, Argentina			
	Nov. 01	08:52	22:23
	Nov. 30	08:34	22:51
Cape Town, South Africa			
	Nov. 01	03:46	17:14
	Nov. 30	03:28	17:42
London, UK			
	Nov. 01	06:55	16:34
	Nov. 30	07:43	15:56
Los Angeles, USA			
	Nov. 01	14:13	01:01
	Nov. 30	14:40	00:44
Nairobi, Kenya			
	Nov. 01	03:12	15:21
	Nov. 30	03:16	15:27
Sydney, Australia			
	Nov. 01	18:54	08:23
	Nov. 30	18:37	08:50
Tokyo, Japan			
	Nov. 01	21:03	07:46
	Nov. 30	21:32	07:28
Washington, DC, USA			
	Nov. 01	11:36	22:08
	Nov. 30	12:07	21:47
Wellington, New Zealand			
	Nov. 01	17:07	07:02
	Nov. 30	16:43	07:36

NB: the times given are in Universal Time (UT)

The Moon's phases and ages

Northern hemisphere

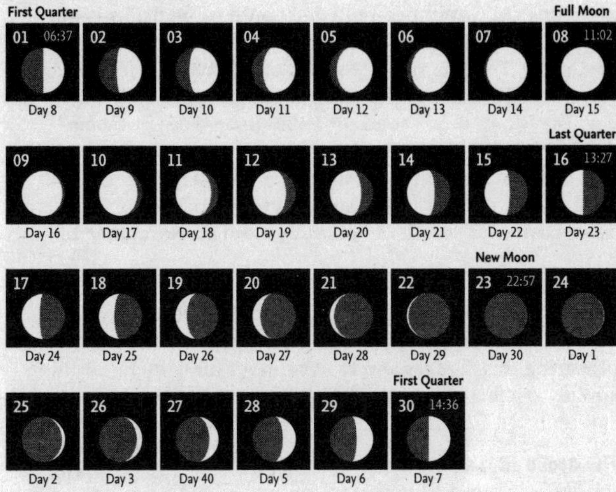

Southern hemisphere

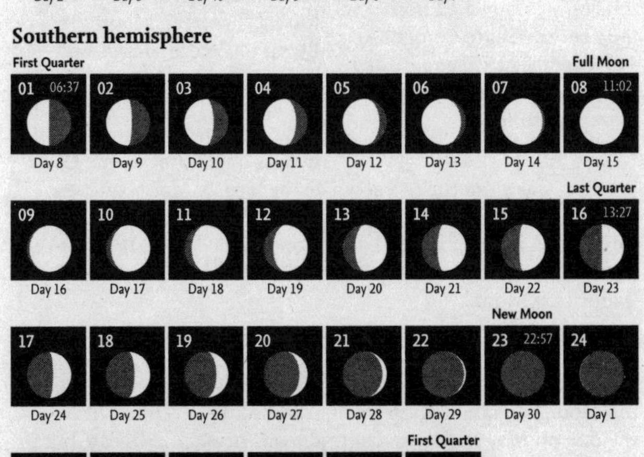

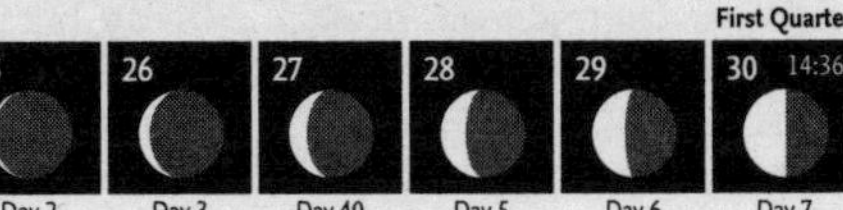

The Moon

Just after First Quarter on November 1, the Moon is 4.2° south of ***Saturn.*** On November 4 it is 2.4° south of ***Jupiter*** in ***Pisces.*** There is a total lunar eclipse at Full Moon on November 8 (see next page). The Moon is 8.0° north of ***Aldebaran*** in ***Taurus*** on November 10 and is 2.5° north of ***Mars*** (also in Taurus) the next day. The Moon is 1.7° south of ***Pollux*** (mag. 1.16) in ***Gemini*** on November 14. Just after Last Quarter, on November 16, the Moon is 5.0° north of ***Regulus*** (α Leonis). On November 21 it is 4.2° north of ***Spica*** (mag. 0.98) in ***Virgo.*** On November 29, it again passes 4.2° south of ***Saturn***, as it did on November 1.

Beaver Moon

The November Full Moon has come to be called 'Beaver Moon', because beavers become particularly active at this time, preparing their lodges and food supplies for winter. But this applies to a few only of the North-American tribes. Most see it as marking the beginning of heavy frosts, with names such as 'Freezing River Maker Moon', 'Freezing Moon', 'Rivers Begin to Freeze Moon', and 'Frost Moon', although to the Haida of Alaska, where snow lies for many months, it was 'Snow Moon'. In the Old World it was sometimes known as the 'Frosty Moon' or even, occasionally, as the 'Oak Moon', although the latter term was more often applied to the Full Moon in December. If it was the last Full Moon before the winter solstice it was also called the 'Mourning Moon'.

The ***Apollo 12*** manned lunar mission was launched from the Kennedy Space Center on 14 November 1969 and landed in Oceanus Procellarum on 19 November 1969.

Total lunar eclipse of November 8

The second lunar eclipse of 2022 is a total eclipse, as none of the Moon remains within the penumbra. This eclipse, like the one on May 16, is also centred over the Pacific Ocean, but this time it is over the northern region, from which the whole eclipse will be visible. The greatest phase, beginning at 09:09, when the Moon starts to enter the Earth's umbra, will be visible from the whole of North America, and last until 11:42 UT (the Moon will then be setting as seen from the West Coast of the United States and Canada). At that time, the eclipsed Moon will be rising over northern Australia, where only the final penumbral phase will be readily detectable. (But, of course, the penumbral phases of any lunar eclipse are effectively invisible to the naked eye, although the change in illumination may be detectable with carefully exposed photographs.)

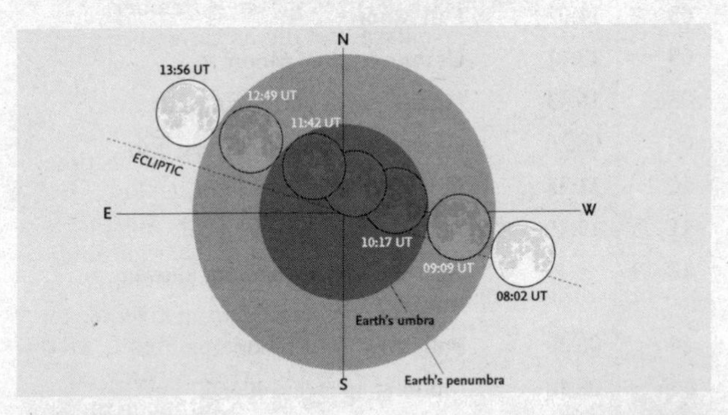

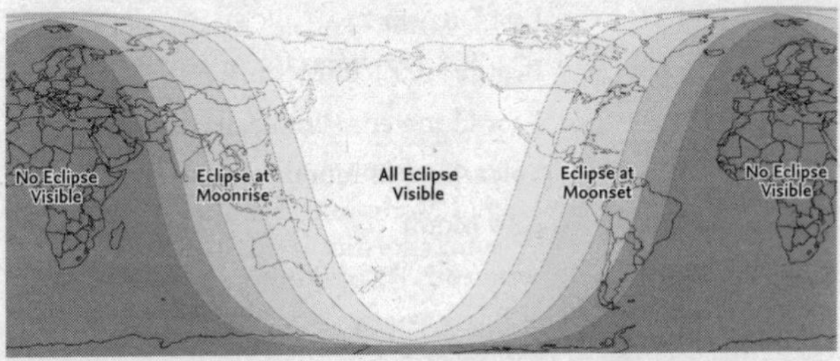

The lines on the map indicate where different phases of the eclipse occur. Starting in the east (right) they show when the Moon first contacts the penumbra; when it fully enters the penumbra; first contact with the dark umbra; and when it fully enters the umbra. The later lines indicate corresponding phases at the end of the eclipse: Moon leaving umbra; Moon fully outside umbra; Moon leaving penumbra.

Calendar for November

01	06:37	First Quarter
01	21:08	Saturn 4.2°N of Moon
04	08:20	Neptune 3.2°N of Moon
04	20:24	Jupiter 2.4°N of Moon
06–30		Leonid meteor shower
08	10:58	Total lunar eclipse
08	11:02	Full Moon
08	13:11	Uranus 0.8°S of Moon
08	16:43	Mercury at superior conjunction
09	08:26	Uranus at opposition (mag. 5.6)
10	11:22	Aldebaran 8.0°S of Moon
11	13:46	Mars 2.5°S of Moon
12–13		Northern Taurid meteor shower maximum
14	00:20	Pollux 1.7°N of Moon
14	06:40	Moon at apogee = 404,921 km
16	13:27	Last Quarter
16	21:53	Regulus 5.0°S of Moon
17–18		Leonid meteor shower maximum
21	04:11	Spica 4.2°S of Moon
23	22:57	New Moon
24	12:18	Antares 2.3°S of Moon
24	14:02	Venus 2.3°N of Moon
24	15:03	Mercury 0.9°N of Moon
26	01 31	Moon at perigee = 362,826 km
28–Dec.09		Phoenicid meteor shower
29	04:40	Saturn 4.2°N of Moon
30	14:36	First Quarter

November 1–4 • *Shortly after First Quarter (on November 1), the Moon passes Saturn. On November 4 it is close to Jupiter. Fomalhaut is close to the southern horizon (as seen from London).*

November 21 • *The Moon with Spica, in the southeast (as seen from London).*

November 28-29 • *The Moon passes below Saturn, almost due south (as seen from central USA).*

N

Early November 23:00 — Mid November 22:00 — Late November 21:00

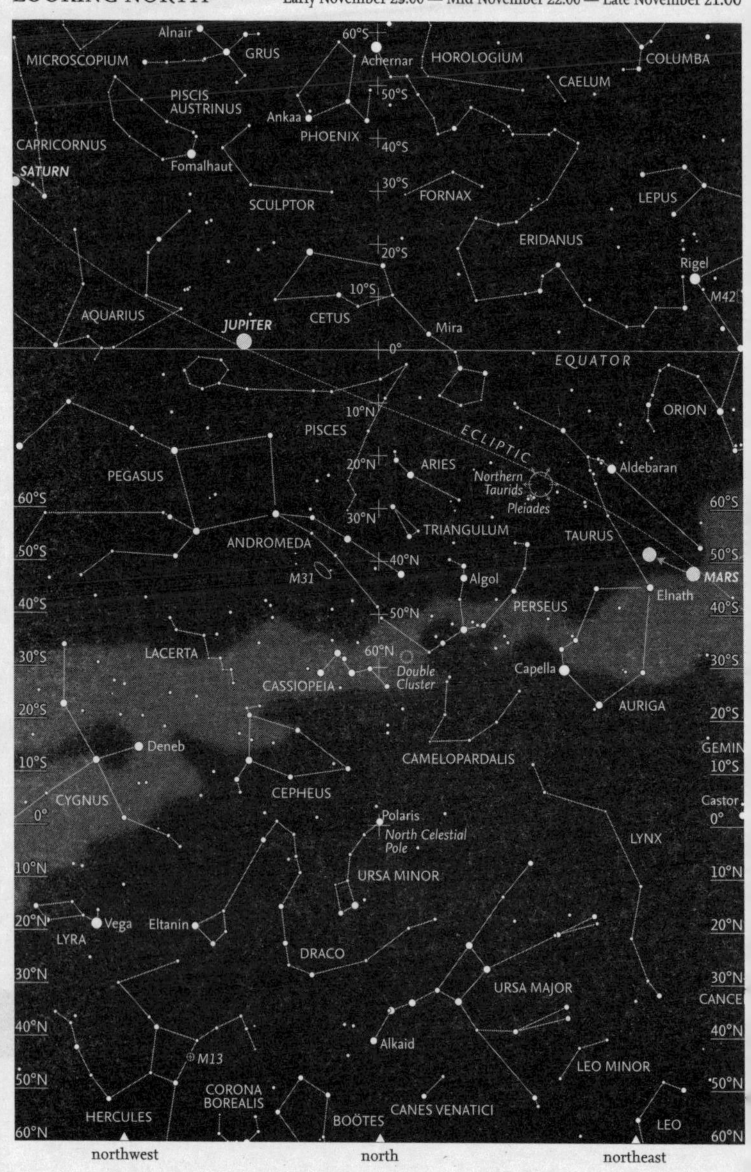

November – Looking North

For observers at mid-northern latitudes, the Milky Way is now high in the north and northwest. ***Cassiopeia*** is not far from the zenith and ***Cepheus*** is 'on its side' in the northwest. ***Capella*** (α Aurigae) and the whole constellation of ***Auriga*** are high in the northeast. One of the stars of the Summer Triangle, Altair in Aquila, is now disappearing below the horizon. The other two: ***Deneb*** in ***Cygnus*** and ***Vega*** in ***Lyra*** are still visible, but low towards the northwestern horizon. ***Eltanin*** (γ Draconis) the brightest star in the 'head' of ***Draco***, is at about the same altitude as Vega. ***Ursa Minor***, together with ***Polaris*** itself, is slightly higher in the sky.

The end of the 'tail' of ***Ursa Major*** (that is, the star ***Alkaid***, η Ursae Majoris), is almost due north, although the body of the constellation has begun to swing round into the northeast. Most of the fainter stars in the constellation to the south and east are easily visible, as is the undistinguished constellation of ***Canes Venatici*** to the southwest of it. Another insignificant constellation, often ignored, is ***Leo Minor*** to the southeast of Ursa Major. This consists of little more than three faint stars. ***Gemini***, with the two bright stars, ***Castor*** and Pollux, is off to the east, with the line of stars forming ***Lynx*** between that constellation and Ursa Major. To the west of Ursa Major is ***Hercules*** although some of its stars are very low, close to the horizon, and likely to be difficult to see. Slightly towards the south are the northernmost stars of ***Boötes***, although Arcturus itself is below the northern horizon.

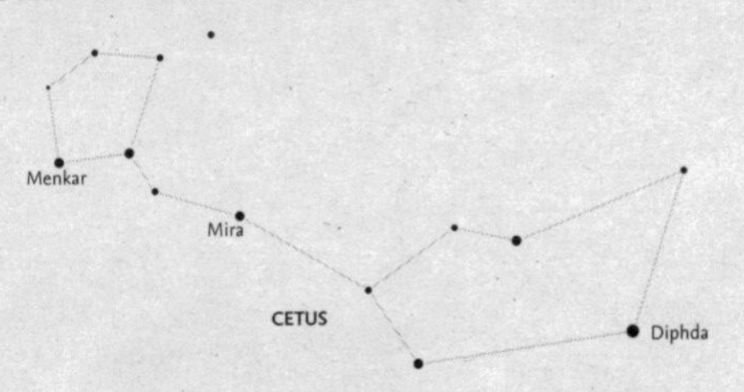

This time of the year, the constellation of Cetus and its famous variable star, Mira (o Ceti), is clearly visible. Because the celestial equator crosses the constellation, it is visible from almost every latitude.

N

Double Cluster
CASSIOPEIA
LACERTA
CYGNUS
AURIGA
PERSEUS
MARS
Elnath
Algol
M31
ANDROMEDA
TRIANGULUM
Pleiades
Northern Taurids
Aldebaran
TAURUS
ARIES
PEGASUS
ECLIPTIC
PISCES
ORION
EQUATOR
JUPITER
Mira
Cursa
M42
Rigel
CETUS
AQUARIUS
ERIDANUS
LEPUS
FORNAX
SCULPTOR
Fomalhaut
SATURN
Acamar
Ankaa
PISCIS AUSTRINUS
CAPRICORNUS
CAELUM
HOROLOGIUM
Achernar
PHOENIX
GRUS
Alnair
DORADO
COLUMBA
MICROSCOPIUM
PICTOR
RETICULUM
TUCANA
INDUS
Canopus
LMC
SMC
HYDRUS
Tarantula N.
MENSA
Peacock
SAGITTARIUS
PUPPIS
South Celestial Pole
PAVO
CARINA
VOLANS
OCTANS
CHAMAELEON
APUS
TRIANGULUM AUSTRALE
TELESCOPIUM
MUSCA
CORONA AUSTRALIS
VELA
Eta Car Neb.
ARA
PYXIS
Acrux
Hadar
CIRCINUS
CRUX
Rigil Kentaurus
Mimosa
NORMA
ANTLIA
SCORPIUS
CENTAURUS
Omega Centauri
LUPUS
MERCURY
Antares
OPHIUCHUS
HYDRA
LIBRA
VENUS
60°N
50°N
40°N
30°N
20°N
10°N
0°
10°S
20°S
30°S
40°S
50°S
60°S
southeast
south
southwest

November – Looking South

Several major constellations dominate the southern sky. There is ***Pegasus*** with the Great Square and ***Andromeda*** above and to the east of it. ***Pisces*** straddles the meridian, with the 'Circlet' to the south of the Great Square. The zodiacal constellation of ***Aquarius*** is sinking in the west, but is still clearly visible. To the east is ***Taurus*** and orange-tinted ***Aldebaran***. Below (south) of Pisces the whole of ***Cetus*** and its famous variable star, ***Mira*** (o Ceti), is clearly visible. ***Orion*** is rising in the east, and the beginning of ***Eridanus*** at the stars λ Eridani and ***Cursa*** (β Eridani) near ***Rigel*** (β Orionis) is visible as are most of the stars in the long chain that forms this constellation as it winds its way south. The brightest star, ***Achernar*** (α Eridani), at the very end of the constellation, is on the horizon for observers at a latitude of 30° north. This constellation once ended at ***Acamar***, the moderately bright star south of the small constellation of ***Fornax***, and originally called Achernar until the constellation was extended and the name transferred to the current, brighter star.

On 3 November 1973, the ***Mariner 10*** spaceprobe, built to study Venus and Mercury, was launched.

Immediately north and west of Achernar is ***Phoenix*** and between that and Aquarius lie the two constellations of ***Sculptor*** and ***Piscis Austrinus***, the latter with its solitary bright star, ***Fomalhaut***. Below Piscis Austrinus is the constellation of ***Grus***. Both the Magellanic Clouds are clearly seen. The Small Cloud ***(SMC)*** on the border between ***Tucana*** and ***Hydrus*** is almost on the meridian, and the Large Cloud ***(LMC)*** is farther east, bordered by ***Dorado*** and ***Mensa***. ***Canopus*** (α Carinae) is at almost the same altitude, and the stars of ***Puppis*** are farther east. South of Canopus are the stars of ***Vela*** and ***Carina***. ***Crux*** is now more-or-less 'upright', just east of the meridian. ***Rigil Kentaurus*** and ***Hadar*** are on the other side of the meridian. All these significant stars are right on the horizon for observers at 30° south, and only visible later in the night and later in the month. The zodiacal constellation of ***Scorpius*** is disappearing in the southwest and the stars of the sprawling constellation of ***Centaurus*** and those of ***Lupus*** are low in the sky.

A supernova that appeared in Cassiopeia on 6 November 1572 (observed by Tycho Brahe) provided firm evidence that change occurred in the heavens. It was last observed on 19 May 1574.

December

December – Introduction

There is one significant meteor shower in December (the last major shower of the year). This is the ***Geminid*** shower, which is visible over the period December 4–20 and comes to maximum on December 14–15, when the Moon is waning gibbous so conditions are not wonderful. It is one of the most active showers of the year, and in some years is the strongest, with a peak rate of around 100 meteors per hour. It is the one major shower that shows good activity before midnight. The meteors have a much higher density than most meteors (which are derived from cometary material). It was eventually established that the Geminids and the asteroid Phaethon had similar orbits. The Geminids are assumed to consist of denser, rocky material from Phaethon. They are slower than most other meteors and often seem to last longer. The brightest frequently break up into numerous luminous fragments that follow similar paths across the sky. There is a second, minor, northern shower: the ***Ursids***, active December 17–26, peaking on December 22–23. The maximum rate is approximately 10 meteors per hour.

The ***OSIRIS-REx*** spaceprobe rendezvoused with minor planet ***(101955) Bennu*** on 3 December 2018.

There are two southern meteor showers that may be seen in December. The ***Phoenicid*** shower begins in late November, and continues into December, reaching its weak maximum on December 2. The other shower, the ***Puppid Velids*** has its radiant on the border between the two constellations of ***Puppis*** and ***Vela***. The shower begins on December 1, lasting until December 15, with maximum on December 7. (In 2022, this is one day before Full Moon.) It is a weak shower with a maximum rate of about 10 meteors per hour, although bright meteors are frequently visible.

The ***Hayabusa 2*** spaceprobe returned its samples of the minor planet ***(162173) Ryugu*** to Woomera in Australia on 6 December 2020.

The planets

Mercury reaches greatest elongation east in the evening sky (20.1°, mag. -0.6) on December 21. ***Venus*** (mag. -3.9) becomes visible, also in the evening sky, later in the month. ***Mars*** is in ***Taurus*** and comes to opposition on December 8 at mag. -1.9, when it is also occulted by the Full Moon. Although conditions will be difficult because of the brilliance of the Moon, the occultation should be visible from Western Europe and practically the whole of North America. ***Jupiter*** (mag. -2.6 to -2.4) is moving eastwards (with direct motion) in ***Pisces***. ***Saturn*** is in ***Capricornus*** at mag. 0.8. ***Uranus*** is still retrograding in ***Aries***, and fades very slightly to mag. 5.7 at the end of the year. ***Neptune*** remains in ***Aquarius*** at mag. 7.6, but resumes direct motion on December 26.

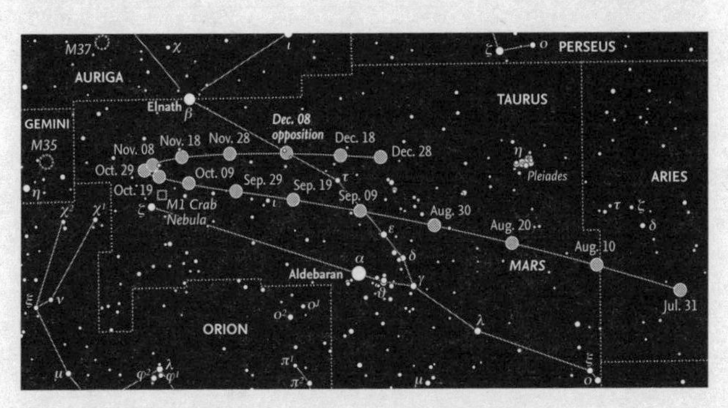

The path of Mars, from July 31 to December 28. Mars comes to opposition on December 8, at magnitude -1.9.

On 9 December 1978, the three small, and the one large probe of the ***Pioneer Venus Multiprobe*** mission entered the atmosphere of Venus.

D

Sunrise and sunset

City	Date	Sunrise	Sunset
Buenos Aires, Argentina			
	Dec. 01	08:34	22:52
	Dec. 31	08:43	23:10
Cape Town, South Africa			
	Dec. 01	03:28	17:42
	Dec. 31	03:38	18:00
London, UK			
	Dec. 01	07:45	15:55
	Dec. 31	08:07	16:01
Los Angeles, USA			
	Dec. 01	14:13	01:01
	Dec. 31	14:40	00:44
Nairobi, Kenya			
	Dec. 01	03:16	15:27
	Dec. 31	03:30	15:42
Sydney, Australia			
	Dec. 01	18:37	08:51
	Dec. 31	18:47	09:09
Tokyo, Japan			
	Dec. 01	21:33	07:28
	Dec. 31	21:51	07:37
Washington, DC, USA			
	Dec. 01	12:08	21:47
	Dec. 31	12:27	21:56
Wellington, New Zealand			
	Dec. 01	16:42	07:37
	Dec. 31	16:51	07:57

NB: the times given are in Universal Time (UT)

The Moon's phases and ages

Northern hemisphere

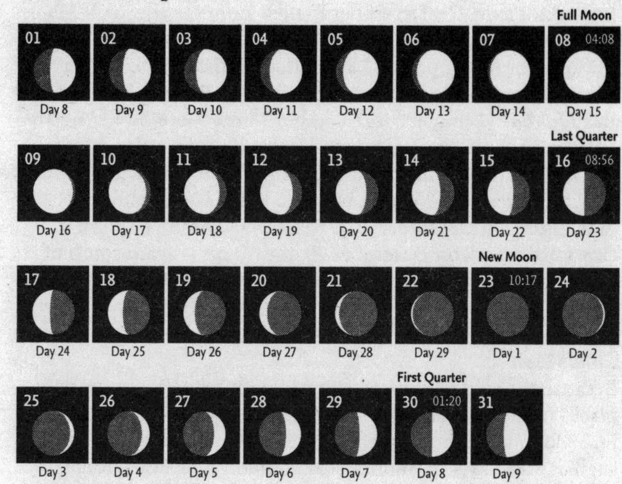

Southern hemisphere

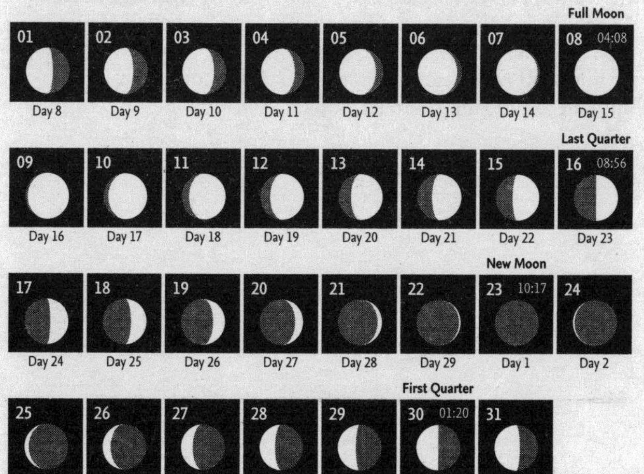

The Moon

On December 2, the waxing gibbous Moon is 2.5° south of ***Jupiter*** in ***Pisces***. On December 8, a few minutes after Full Moon, it occults ***Mars*** (which is at opposition) in ***Taurus***. When waning gibbous, the Moon passes 1.8° south of ***Pollux*** on December 11. On December 14 it passes 4.8° north of ***Regulus***, between it and ***Algieba*** (γ Leonis). On December 18 the Moon is 4.1° north of ***Spica*** in ***Virgo***. On December 21, shortly before New Moon, it is 2.3° north of ***Antares***. On December 24 it passes just south of both Mercury and Venus, but is so low it is probably lost in twilight. On December 26 it passes 4.0° south of ***Saturn*** in ***Capricornus*** and, on December 29, the Moon is again south of Jupiter as it was early in the month, but this time slightly closer at a distance of 2.3°.

Cold Moon

Because the cold of winter begins to dominate life for most places in the northern hemisphere, many of the names for the Full Moon in December refer to the temperature, with terms such as 'Cold Moon', 'Winter Maker Moon' and 'Snow Moon'. The Cree of Canada had a rather strange name: 'Moon when the Young Fellow Spreads the Brush'. Among the Zuñi of New Mexico it was the 'Moon when the Sun has Travelled Home to Rest'. In Europe it was sometimes called the 'Moon before Yule' or the 'Wolf Moon', although that term was more commonly applied to the Full Moon in January.

South Pole–Aitken Basin
On 3 January 2019, the Chinese *Chang'e-4* probe soft-landed in Von Kármán crater, within the South Pole–Aitken Basin on the lunar far side. The Von Kármán crater is some 180 km across, and the northern portion is buried beneath ejecta from the larger Leibnitz crater.

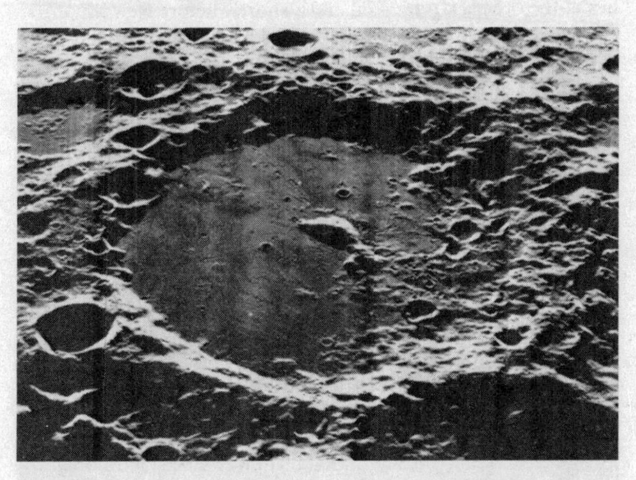

This image of the Von Kármán crater was obtained by the Lunar Orbiter 5 *spaceprobe during its operation between 6 August and 18 August 1967. The view is looking west. The rough terrain in the northern portion has been created by ejecta from the Leibnitz crater.*

The giant South Pole–Aitken Basin is the largest impact structure on the Moon. At one time it was considered the largest impact feature in the Solar System, but it is now known that Valhalla, on Jupiter's satellite Callisto, is larger. Valhalla is about 3800 km in diameter, whereas the South Pole–Aitken Basin is about 2500 km across.

Because of its location, little of the Basin is visible from Earth, even under favourable libration conditions. Occasionally, part of the outermost ring of mountains may be seen at the

Moon's southern pole, where they are known informally as the Leibnitz Mountains. The peaks of these mountains are the highest on the Moon, being about 8000 m (8 km) above the reference level.

The name of the basin arises from the fact that the lunar South Pole is at one edge of the feature, and the crater Aitken lies at the other extreme.

The approximate location of the outer ring of mountains around the Moon's South Pole–Aitken Basin. The South Pole itself is not visible in this image. It actually lies on this outer ring.

Although the existence of a large basin was suspected back in 1962, it was not until mapping was carried out by the various Lunar Orbiter missions that the great size of the basin was appreciated. The full size and depth of the basin was determined from data obtained by the ***Clementine*** mission. Just as the highest peaks are part of the outer rim, so the lowest elevations on the Moon (about 6000 m below the reference level) occur within the basin. It has been established that the crust beneath the basin is about 30 km in thickness. This compares with some 60–80 km in its surroundings and an overall crustal thickness of about 50 km.

It has been found that the surface of the basin has a different composition to that of the surrounding highlands. In particular, there are slightly higher concentrations of three elements: iron, titanium and thorium. Similarly, two specific minerals, clinopyroxene and orthopyroxene, are more abundant. This difference in composition appears, even visually, in images of the basin. It appears darker than its surroundings (see, for example, the image on page 227 and the image of the lunar far side on page 100). There are various suggestions as to how this difference in composition has come about, and only a sample return will enable the true explanation to be obtained. In May 2019 it was reported that the Chinese *Chang-e 4* lander had observed mantle rocks at the surface within Von Kármán crater. This would imply that the impactor that formed the basin itself did penetrate to a considerable depth, bringing mantle material to the surface. (The impact that created the Von Kármán crater that was the actual landing site would not have been sufficiently energetic to penetrate to the mantle.)

On 25 December 2004, the ***Huygens*** probe, which would land on Titan in January 2005, separated from the ***Cassini*** Saturn orbiter.

D

Calendar for December

01–15		Puppid Velid meteor shower
01	13:22	Neptune 3.2°N of Moon
02		Phoenicid meteor shower maximum
02	00:57	Jupiter 2.5°N of Moon
04–20		Geminid meteor shower
05	17:59	Uranus 0.7°S of Moon
07		Puppid Velid meteor shower maximum
07	18:43	Aldebaran 8.0°S of Moon
08	04:08	Full Moon
08	04:25	Occultation of Mars
08	05:42	Mars at opposition (mag. -1.9)
11	07:44	Pollux 1.8°N of Moon
12	00:28	Moon at apogee = 405,869 km
14–15		Geminid meteor shower maximum
14	05:38	Regulus 4.8°S of Moon
16	08:56	Last Quarter
17–26		Ursid meteor shower
18	14:13	Spica 4.1°S of Moon
21	15:31	Mercury at greatest elongation (20.1°E, mag. -0.6)
21	21:48	December solstice
21	23:15	Antares 2.3°S of Moon
22	04:00 *	Mars 8.2°N of Aldebaran
22–23		Ursid meteor shower maximum
23	10:17	New Moon
24	08:27	Moon at perigee = 358,270 km
24	11:28	Venus 3.8°N of Moon
24	18:30	Mercury 3.8°N of Moon
26	16:11	Saturn 4.0°N of Moon
28	20:03	Neptune 3.0°N of Moon
29	09:00 *	Mercury 1.4°N of Venus
29	10:34	Jupiter 2.3°N of Moon
30	01:21	First Quarter

** These objects are close together for an extended period around this time.*

December 8 • *The Full Moon with Mars, shortly before the occultation at 04:25 (as seen from London).*

December 24 • *Low in the southwest, the narrow crescent Moon is close to Mercury and Venus (as seen from central USA).*

December 29 • *Venus and Mercury, shortly after sunset. Mercury is much fainter than Venus (as seen from central USA).*

December 29 • *The Moon and Jupiter are close together in the western sky (as seen from Sydney).*

D

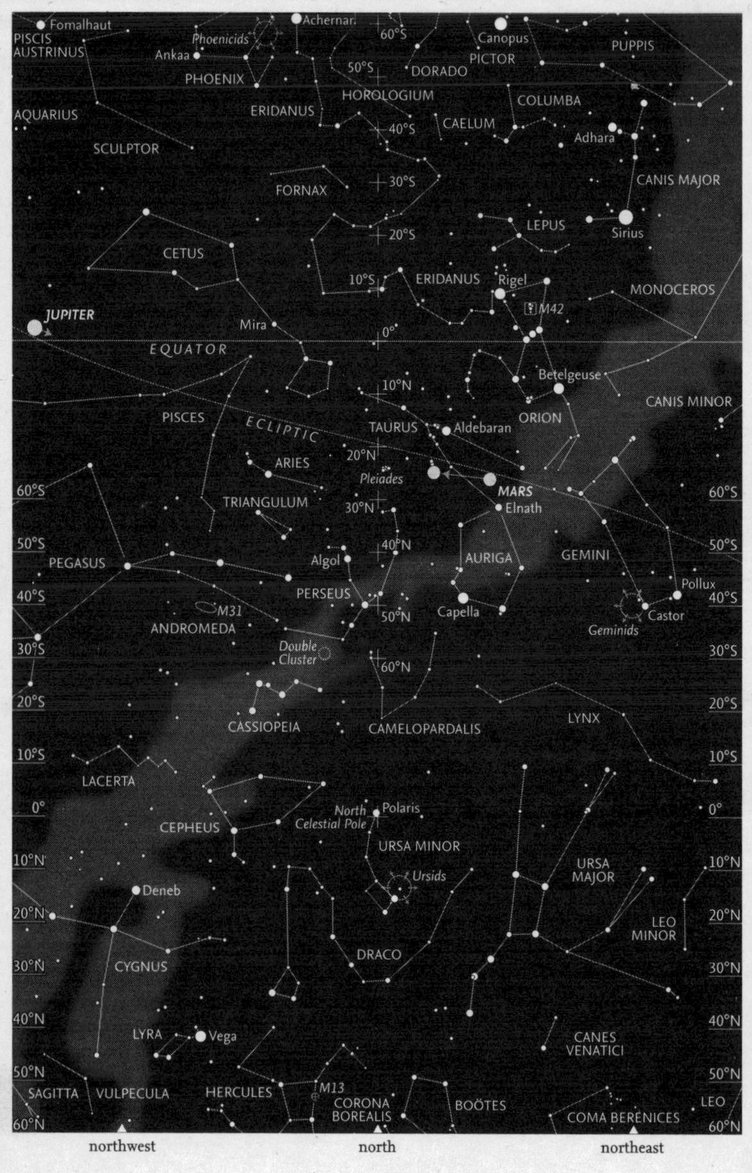
Fomalhaut
PISCIS AUSTRINUS
Phoenicids
Achernar
60°S
Canopus
PUPPIS
Ankaa
PICTOR
PHOENIX
50°S
DORADO
HOROLOGIUM
COLUMBA
AQUARIUS
ERIDANUS
40°S
CAELUM
Adhara
SCULPTOR
30°S
CANIS MAJOR
FORNAX
LEPUS
20°S
Sirius
CETUS
10°S
ERIDANUS
Rigel
MONOCEROS
M42
JUPITER
Mira
0°
EQUATOR
Betelgeuse
10°N
CANIS MINOR
PISCES
ORION
ECLIPTIC
TAURUS
Aldebaran
ARIES
20°N
Pleiades
MARS
TRIANGULUM
30°N
Elnath
40°N
GEMINI
PEGASUS
Algol
AURIGA
PERSEUS
Pollux
M31
50°N
Capella
Castor
ANDROMEDA
Geminids
Double Cluster
60°N
CASSIOPEIA
CAMELOPARDALIS
LYNX
LACERTA
North Celestial Pole
Polaris
CEPHEUS
URSA MINOR
URSA MAJOR
Ursids
Deneb
LEO MINOR
DRACO
CYGNUS
LYRA
Vega
CANES VENATICI
SAGITTA
VULPECULA
HERCULES
M13
CORONA BOREALIS
BOÖTES
COMA BERENICES
LEO
60°S
50°S
40°S
30°S
20°S
10°S
0°
10°N
20°N
30°N
40°N
50°N
60°N
northwest
north
northeast

December – Looking North

For observers close to the equator, ***Orion*** is high overhead, to the east of the meridian, with ***Taurus***, ***Auriga*** and ***Gemini*** way above the northern horizon. For observers at mid-northern latitudes, it is ***Perseus*** that is at the zenith, with ***Andromeda*** stretching off to the west and the Great Square of ***Pegasus*** in the northwest. Auriga with ***Capella*** is slightly to the east. Even farther east are the two bright stars of ***Gemini***, ***Castor*** and ***Pollux***.

The Milky Way runs from high in the northeast down to the northwest. Below Perseus is the distinctive shape of ***Cassiopeia***, which lies within the Milky Way, and farther down is the zig-zag constellation of ***Lacerta***, which is like ***Cepheus***, in that both of them are partly within the band of stars. Even farther towards the northwest is ***Cygnus*** and the beginning of the Great Dark Rift near ***Deneb***. The constellation of ***Lyra***, with brilliant ***Vega***, lies towards the meridian, away from the star clouds of the Milky Way. Much of the constellation of ***Hercules*** is clear of the northern horizon, together with some of the northernmost stars in ***Boötes***.

Ursa Major is climbing in the northeast, and all the far-flung outlying stars are clearly visible. Below the 'tail' is the inconspicuous constellation of ***Canes Venatici***, and some observers at high latitudes may even be able to glimpse some of the stars of ***Coma Berenices***, low on the horizon in the northeast. Between Ursa Major and the Milky Way, the whole of the constellations of ***Draco*** and ***Ursa Minor*** are clearly visible as is Cepheus beyond them. The chain of stars forming the faint constellation of ***Lynx*** lies between Ursa Major and the bright stars of Gemini.

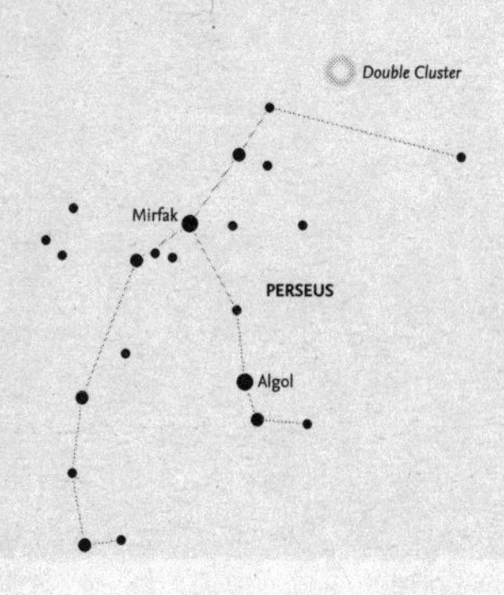

This time of the year, the constellation of Perseus, with the famous variable star Algol and the Double Cluster, is at the meridian.

D

Early December 23:00 — Mid December 22:00 — Late December 21:00

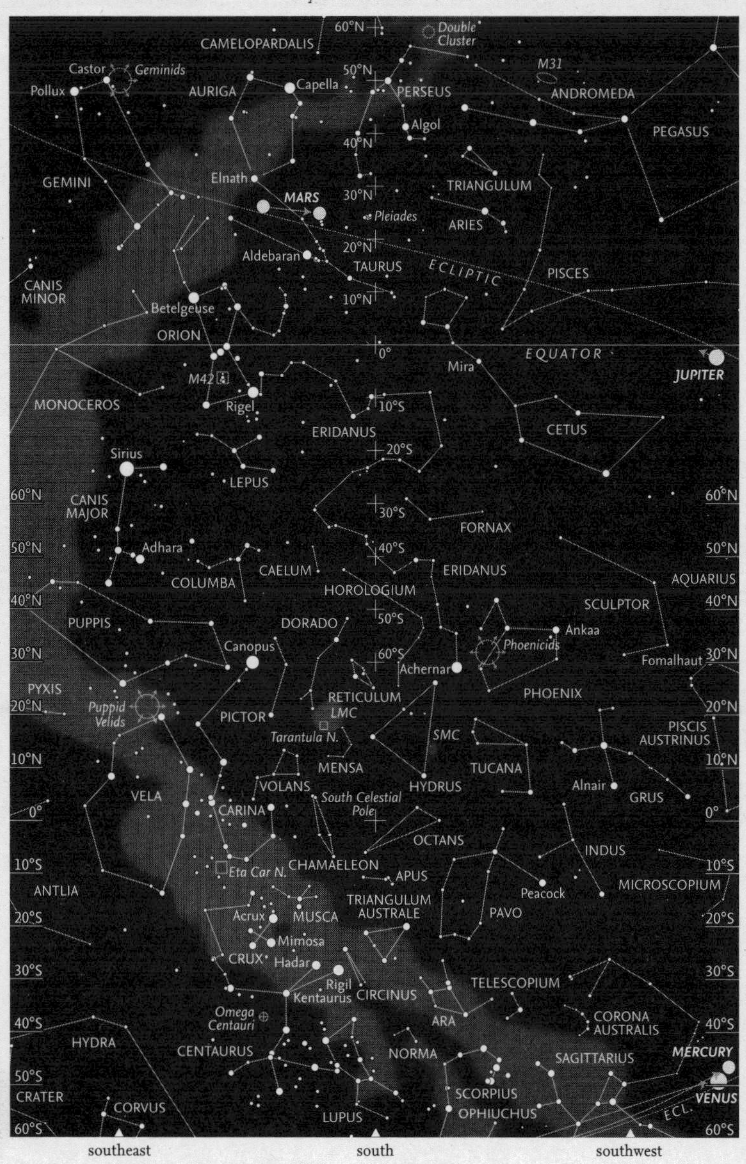

December – Looking South

Orion has now risen well clear of the horizon and is in the northeast. Above it are ***Taurus*** and ***Auriga***, with ***Gemini*** farther to the east. Within the Milky Way to the east of Orion, is the rather undistinguished constellation of ***Monoceros***, which is without any distinct, bright stars. Below Orion is the small constellation of ***Lepus***, and to its east, ***Canis Major***, with brilliant ***Sirius***, the brightest star in the sky (mag. -1.4). The long, winding constellation of ***Eridanus*** begins near ***Rigel*** in Orion and runs south, partly enclosing ***Fornax*** (which was once part of the larger constellation), until it ends at ***Achernar*** (α Eridani). Between Eridanus and Canis Major is the tiny, faint constellation of ***Caelum*** and the larger and brighter ***Columba***.

South of Canis Major is the constellation of ***Puppis***, once (with ***Vela*** and ***Carina***) part of the large obsolete constellation of Argo Navis. ***Canopus*** (α Carinae) and Achernar are close to the horizon for observers at 30°N. West of Achernar is ***Phoenix*** and even farther west, below ***Aquarius***, the constellation of ***Piscis Austrinus*** is slowly descending in the west. Between Piscis Austrinus in the west and Carina in the east lie several small constellations, the most conspicuous of which is ***Grus***. There are also ***Tucana*** and ***Hydrus***, with the ***Small Magellanic Cloud*** (SMC), then the ***Large Magellanic Cloud*** (LMC) in ***Mensa*** and ***Dorado***, with the tiny constellation of ***Reticulum*** north of them. North of that again is the faint constellation of ***Horologium***. Between Mensa and the Milky Way in Carina is the tiny constellation of ***Volans***.

More faint constellations lie around the South Celestial Pole in ***Octans***, notably ***Chamaeleon*** and ***Apus***. To the west lies the larger ***Pavo*** and the rather undistinguished constellation of ***Indus***. ***Crux*** has now swung round into the southeast. Both it and the small neighbouring constellation of ***Musca*** lie in the Milky Way. ***Rigil Kentaurus*** and ***Hadar***, the two brightest stars in ***Centaurus*** are slightly farther south, with ***Circinus*** and ***Triangulum Australe*** between them and Pavo. Those in the far south are able to see all the stars in Centaurus and ***Lupus***, together with the southernmost stars of ***Scorpius*** (the 'sting') and ***Sagittarius***, as well as other constellations, such as ***Norma***, ***Ara***, ***Telescopium*** and ***Corona Australis***.

December: Looking South

[illegible]

Dark Sky Sites

[illegible]

Dark Sky
Sites

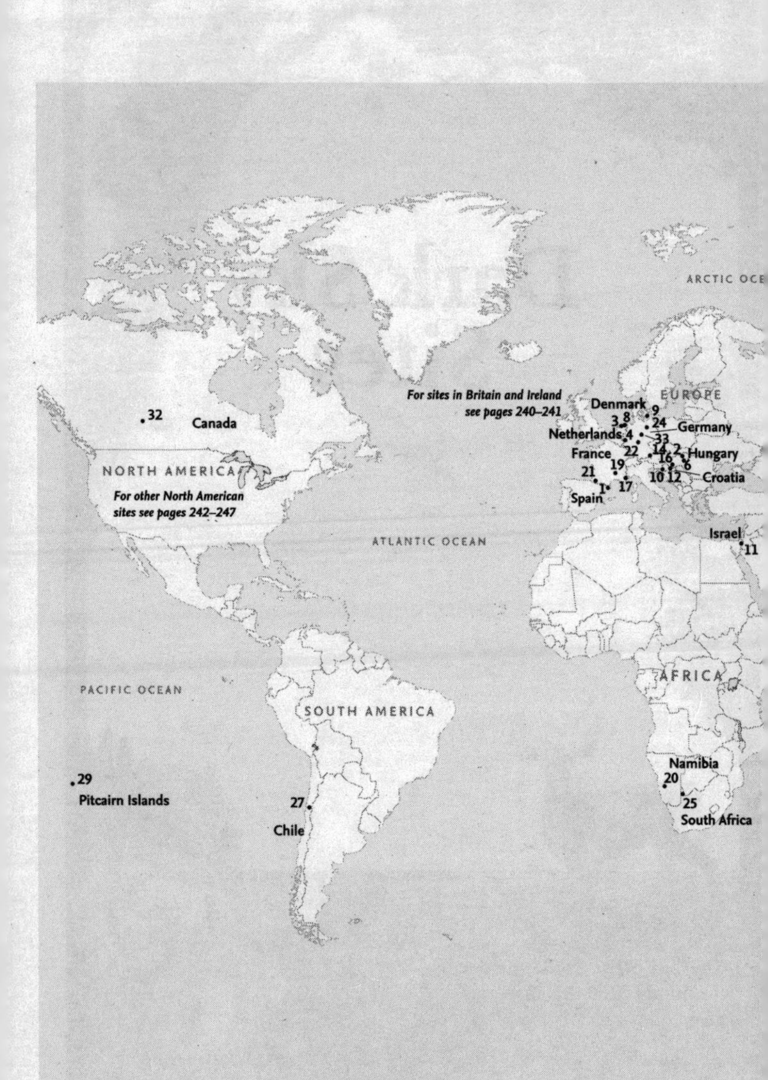
ARCTIC OCE
32
Canada
NORTH AMERICA
For other North American sites see pages 242–247
For sites in Britain and Ireland see pages 240–241
EUROPE
Denmark
9
3
8
24
Germany
Netherlands
4
33
France
22
14
2
Hungary
16
6
19
21
10
12
Croatia
1
17
Spain
ATLANTIC OCEAN
Israel
11
AFRICA
PACIFIC OCEAN
SOUTH AMERICA
29
Pitcairn Islands
27
Chile
Namibia
20
25
South Africa

The 'rest of the world' site details for numbers on this map are given on pages 248–249.

Dark Sky Sites, Britain & Ireland

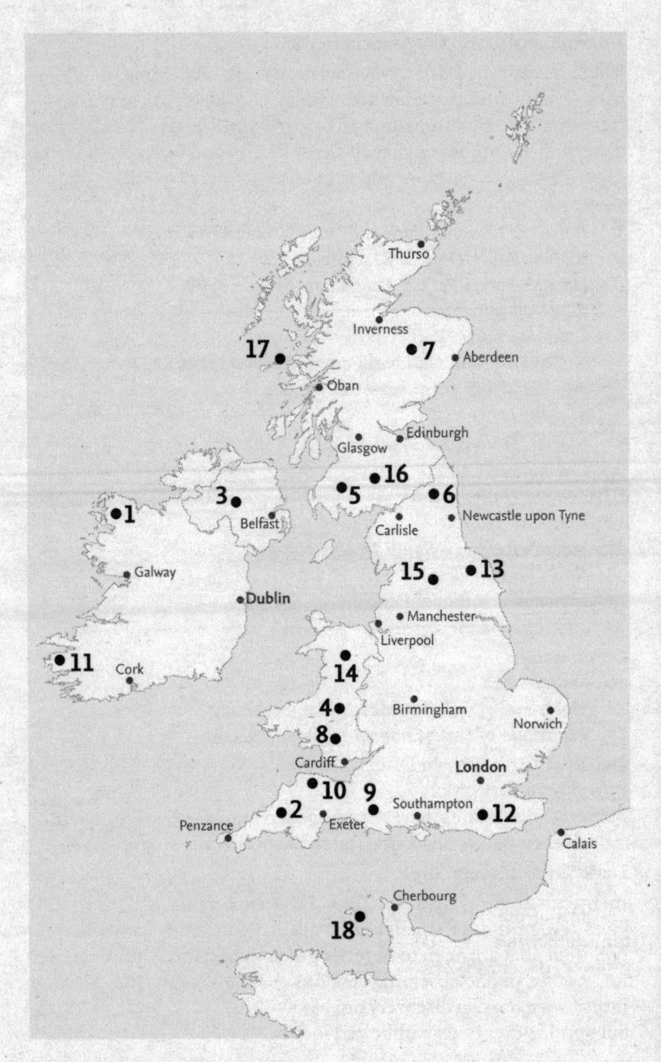

International Dark-Sky Association Sites

The International Dark-Sky Association (IDA) recognizes various categories of sites that offer areas where the sky is dark at night, free from light pollution and particularly suitable for astronomical observing. A number of sites in Great Britain and Ireland have been given specific recognition and are shown on the map. These are:

Parks

1 ***Ballycroy National Park and Wild Nephin Wilderness***
2 ***Bodmin Moor Dark Sky Landscape***
3 ***Davagh Forest Park & Beaghmore Stone Circles***
4 ***Elan Valley Estate***
5 ***Galloway Forest Park***
6 ***Northumberland National Park and Kielder Water & Forest Park***
7 ***Tomintoul and Glenlivet – Cairngorms***

Reserves

8 ***Brecon Beacons National Park***
9 ***Cranbourne Chase***
10 ***Exmoor National Park***
11 ***Kerry***
12 ***Moore's Reserve – South Downs National Park***
13 ***North York Moors***
14 ***Snowdonia National Park***
15 ***Yorkshire Dales***

Communities

16 ***Moffat***
17 ***The island of Coll (Inner Hebrides, Scotland)***
18 ***The island of Sark (Channel Islands)***

Details of these sites and web links may be found at the IDA website:
https://www.darksky.org/

Many of these sites have major observatories or other facilities available for public observing (often at specific dates or times).

Dark Sky Discovery Sites

In Britain there is also the ***Dark Sky Discovery*** organisation. This gives recognition to smaller sites, again free from immediate light pollution, that are open to observing at any time. Some sites are used for specific, public observing sessions. A full listing of sites is at:
https://www.darkskydiscovery.org.uk/
but specific events are publicized locally.

US Dark Sky Sites

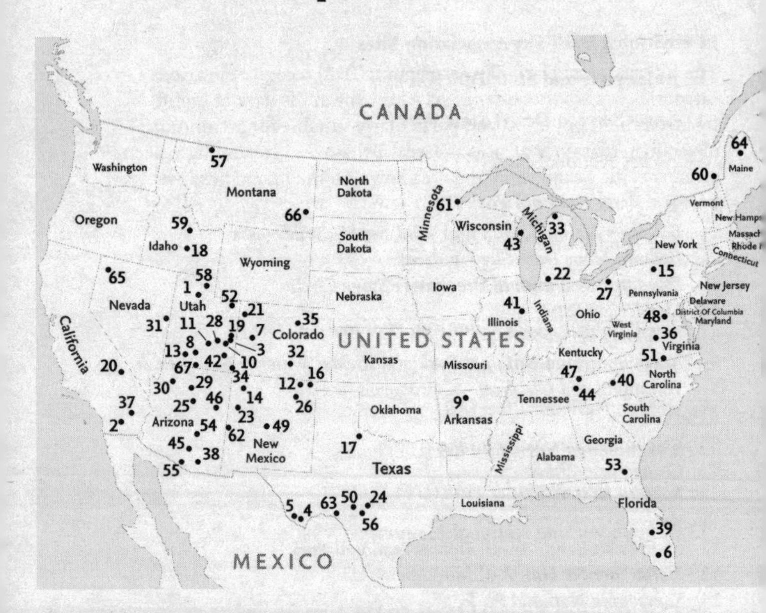

International Dark-Sky Association Sites

The International Dark-Sky Association (IDA) recognises various categories of sites that offer areas where the sky is dark at night, free from light pollution and particularly suitable for astronomical observing. There are numerous sites in North America, shown on the map and listed here. Details of the IDA are at: https://www.darksky.org/.

Information on the various categories and individual sites are at: https://www.darksky.org/our-work/conservation/idsp/

Many of these sites have major observatories or other facilities available for public observing (often at specific dates or times).

Parks

1 ***Antelope Island State Park*** (UT)

2 ***Anza-Borrego Desert State Park*** (CA)

3 ***Arches National Park*** (UT)

4 ***Big Bend National Park*** (TX)

5 ***Big Bend Ranch State Park*** (TX)

6 ***Big Cypress National Preserve*** (FL)

7 ***Black Canyon of the Gunnison National Park*** (CO)

8 ***Bryce Canyon National Park*** (UT)

9 ***Buffalo National River*** (AR)

10 ***Canyonlands National Park*** (UT)

11 ***Capitol Reef National Park*** (UT)

12 ***Capulin Volcano National Monument*** (NM)

13 ***Cedar Breaks National Monument*** (UT)

14 ***Chaco Culture National Historical*** Park (NM)

15 ***Cherry Springs State Park*** (PA)

16 ***Clayton Lake State Park*** (NM)

17 ***Copper Breaks State Park*** (TX)

18 ***Craters Of The Moon National Monument*** (ID)

19 ***Dead Horse Point State Park*** (UT)

20 ***Death Valley National Park*** (CA)

21 ***Dinosaur National Monument*** (CO)

22 ***Dr. T.K. Lawless County Park*** (MI)

23 ***El Morro National Monument*** (NM)

24 ***Enchanted Rock State Natural Area*** (TX)

25 ***Flagstaff Area National Monuments*** (AZ)

26 ***Fort Union National Monument*** (NM)

27 ***Geauga Observatory Park*** (OH)

28 ***Goblin Valley State Park*** (UT)

29 ***Grand Canyon National Park*** (AZ)

30 ***Grand Canyon-Parashant National Monument*** (AZ)

31 ***Great Basin National Park*** (NV)

32 ***Great Sand Dunes National Park and Preserve*** (CO)

33 ***Headlands*** (MI)

34 ***Hovenweep National Monument*** (UT)

35 ***Jackson Lake State Park*** (CO)

36 ***James River State Park*** (VA)

37 ***Joshua Tree National Park*** (CA)

38 ***Kartchner Caverns State Park*** (AZ)

39 ***Kissimmee Prairie Preserve State Park*** (FL)

40 ***Mayland Earth to Sky Park & Bare Dark Sky Observatory*** (NC)

41 ***Middle Fork River Forest Preserve*** (IL)

42 ***Natural Bridges National Monument*** (UT)

43 ***Newport State Park*** (WI)

44 ***Obed Wild and Scenic River*** (TN)

45 ***Oracle State Park*** (AZ)

46 ***Petrified Forest National Park*** (AZ)

47 ***Pickett CCC Memorial State Park & Pogue Creek Canyon State Natural Area*** (TN)

48 ***Rappahannock County Park*** (VA)

49 ***Salinas Pueblo Missions National Monument*** (NM)

50 ***South Llano River State Park*** (TX)

51 ***Staunton River State Park*** (VA)

52 ***Steinaker State Park*** (UT)

53 ***Stephen C. Foster State Park*** (GA)

54 ***Tonto National Monument*** (AZ)

55 ***Tumacácori National Historical Park*** (AZ)

56 ***UBarU Camp and Retreat Center*** (TX)

57 ***Waterton-Glacier International Peace Park*** (Canada/MT)

58 ***Weber County North Fork Park*** (UT)

Reserves

59 ***Central Idaho*** (ID)

60 ***Mont-Mégantic*** (Québec)

Sanctuaries

61 ***Boundary Craters Canoe Area Wilderness*** (MN)

62 ***Cosmic Campground*** (NM)

63 ***Devils River State Natural Area – Del Norte Unit*** (TX)

64 ***Katahdin Woods and Waters National Monument*** (ME)

65 ***Massacre Rim*** (NV)

66 ***Medicine Rocks State Park*** (MT)

67 ***Rainbow Bridge National Monument*** (UT)

RASC Recognized Dark-Sky Sites

Canadian Dark-Sky Sites

The Royal Astronomical Society of Canada (RASC) has developed formal guidelines and requirements for three types of light-restricted protected areas: Dark-Sky Preserves, Urban Star Parks and Nocturnal Preserves. The focus of the Canadian Program is primarily to protect the nocturnal environment; therefore, the outdoor lighting requirements are the most stringent, but also the most effective. Canadian Parks and other areas that meet these guidelines and successfully apply for one of these designations are officially recognized. Many parks across Canada have been designated in recent years – see the list below and the RASC website: https://www.rasc.ca/dark-sky-site-designations.

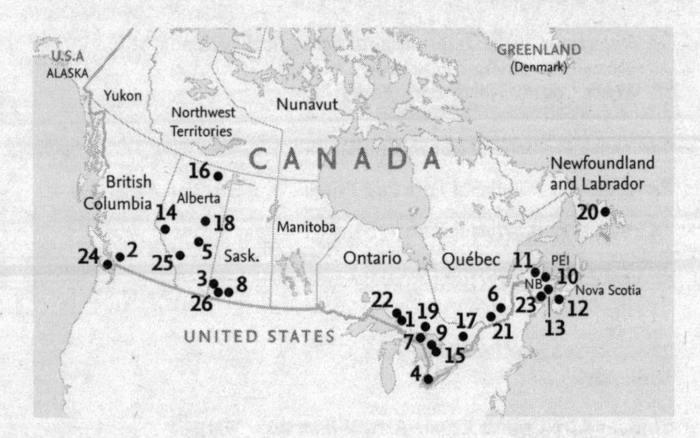

Dark-Sky Preserves

1 ***Torrance Barrens Dark-Sky Preserve*** (ON)

2 ***McDonald Park Dark-Sky Park*** (BC)

3 ***Cypress Hills Inter-Provincial Park Dark-Sky Preserve*** (SK/AB)

4 ***Point Pelee National Park*** (ON)

5 ***Beaver Hills and Elk Island National Park*** (AB)

6 ***Mont-Mégantic International Dark-Sky Preserve*** (QC)

7 ***Gordon's Park*** (ON)

8 ***Grasslands National Park*** (SK)

9 ***Bruce Peninsula National Park*** (ON)

10 ***Kouchibouguac National Park*** (NB)

11 ***Mount Carleton Provincial Park*** (NB)

12 ***Kejimkujik National Park*** (NS)

13 ***Fundy National Park*** (NB)

14 ***Jasper National Park Dark-Sky Preserve*** (AB)

15 ***Bluewater Outdoor Education Centre – Wiarton*** (ON)

16 ***Wood Buffalo National Park*** (AB)

17 ***North Frontenac Township*** (ON)

18 ***Lakeland Provincial Park and Provincial Recreation Area*** (AB)

19 ***Killarney Provincial Park*** (ON)

20 ***Terra Nova National Park*** (NL)

21 ***Au Diable Vert*** (QC)

22 ***Lake Superior Provincial Park*** (ON)

Urban Star Parks

23 ***Irving Nature Park*** (NB)

24 ***Cattle Point, Victoria*** (BC)

Nocturnal Preserves

25 ***Ann and Sandy Cross Conservation Area*** (AB)

26 ***Old Man on His Back Ranch*** (SK)

Dark Sky Parks – World

1 ***Albanyà (Spain)***
2 ***Bükk National Park (Hungary)***
3 ***De Boschplatt (Netherlands)***
4 ***Eifel National Park (Germany)***
5 ***Hehuan Mountain (Taiwan)***
6 ***Hortobágy National Park (Hungary)***
7 ***Iriomote-Ishigaki National Park (Japan)***
8 ***Lauwersmeer National Park (Netherlands)***
9 ***Møn and Nyord (Denmark)***
10 ***Petrova gora-Biljeg (Croatia)***
11 ***Ramon Crater (Israel)***
12 ***Vrani kamen (Croatia)***
13 ***Warrumbungle National Park (Australia)***
14 ***Winklmoosalm (Germany)***
15 ***Yeongyang Firefly Eco Park (South Korea)***
16 ***Zselic National Landscape Protection Area (Hungary)***

Dark Sky Reserves

17 *Alpes Azur Mercantour (France)*

18 *Aoraki Mackenzie (New Zealand)*

19 *Cévennes National Park (France)*

20 *NambiRand Nature Reserve (Namibia)*

21 *Pic du Midi (France)*

22 *Rhön (Germany)*

23 *River Murray (Australia)*

24 *Westhavelland (Germany)*

Dark Sky Sanctuaries

25 *!Ae!Hai Kalahari Heritage Park (South Africa)*

26 *Aotea / Great Barrier Island (New Zealand)*

27 *Gabriela Mistral (Chile)*

28 *Niue (New Zealand)*

29 *Pitcairn Islands (UK)*

30 *Stewart Island / Rakiura (New Zealand)*

31 *The Jump-Up (Australia)*

Dark Sky Communities

32 *Bon Accord (Canada)*

33 *Fulda (Germany)*

Twilight Diagrams

Sunrise, sunset, twilight

For each individual month, we give details of sunrise and sunset times for nine cities across the world. But observing the stars is also affected by twilight, and this varies considerably from place to place. During the summer, especially at high latitudes, twilight may persist throughout the night and make it difficult to see the faintest stars. Beyond the Arctic and Antarctic Circles, of course, the Sun does not set for 24 hours at least once during the summer (and rise for 24 hours at least once during the winter). Even when the Sun does dip below the horizon at high latitudes, bright twilight persists throughout the night, so observing the stars is impossible.

There are three recognized stages of twilight: civil twilight, when the Sun is less than 6° below the horizon; nautical twilight, when the Sun is between 6° and 12° below the horizon; and astronomical twilight, when the Sun is between 12° and 18° below the horizon. Full darkness occurs only when the Sun is more than 18° below the horizon. During nautical twilight, only the very brightest stars are visible. (These are the stars that were used for navigation, hence the name for this stage.) During astronomical twilight, the faintest stars visible to the naked eye may be seen directly overhead, but are lost at lower altitudes. They become visible only once it is fully dark. The diagrams show the duration of twilight at the various cities. Of the locations shown, during the summer months there is full darkness at most of the cities, but it never occurs during the summer at the latitude of London. Observing conditions are most favourable at somewhere like Nairobi, which is very close to the equator, so there is not only little twilight, and a long period of full darkness, but there are also only slight variations in timing and duration throughout the year.

The diagrams also show the times of New and Full Moon (black and white symbols, respectively). As may be seen, at most locations during the year roughly half of New and Full Moon phases may come during daylight. For this reason, the exact phase may be invisible at one location, but be clearly seen elsewhere. The exact times of the events are given in the diagrams for each individual month.

Buenos Aires, Argentina – Latitude 34.7°S – Longitude 58.5°W

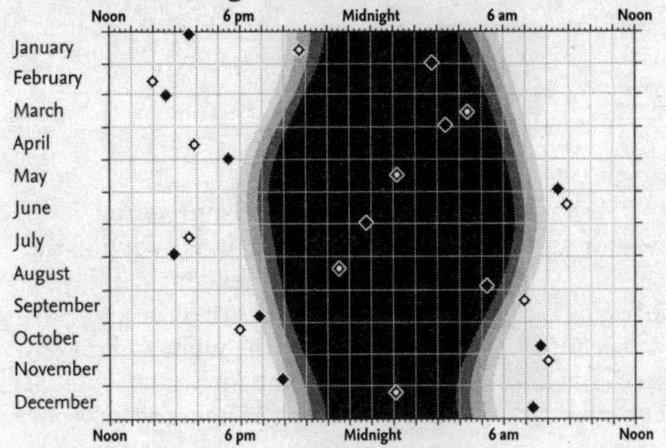

Cape Town, South Africa – Latitude 33.9°S – Longitude 18.5°E

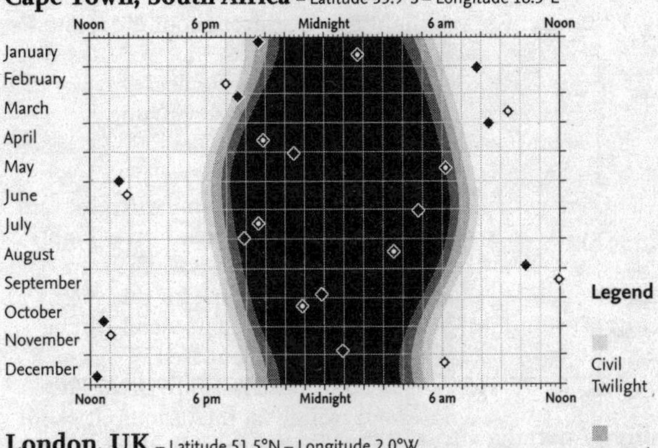

London, UK – Latitude 51.5°N – Longitude 2.0°W

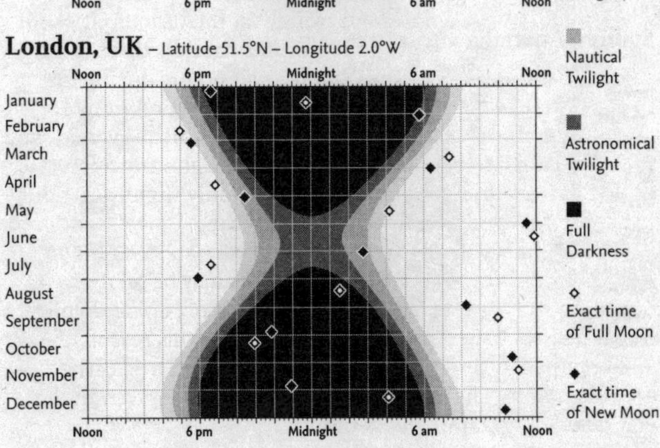

Los Angeles, USA – Latitude 34.0°N – Longitude 118.2° W

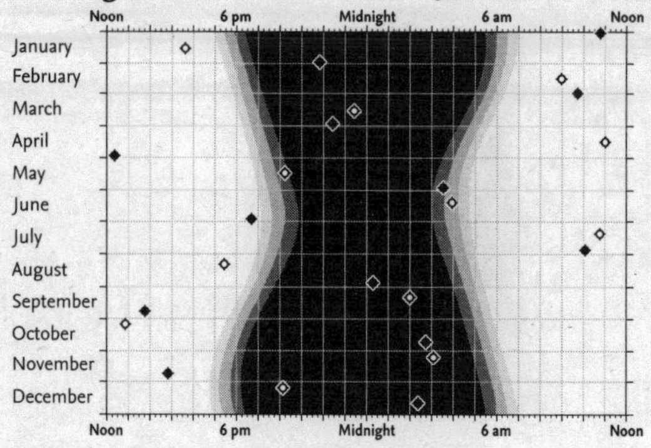

Nairobi, Kenya – Latitude 1.3°S – Longitude 36.8°E

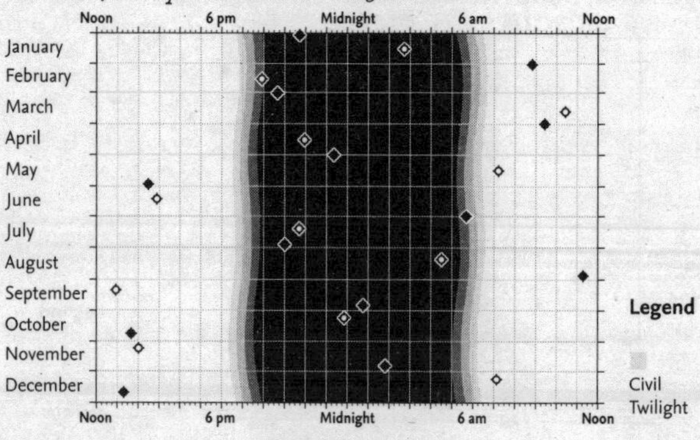

Sydney, Australia – Latitude 33.5°S – Longitude 151.2°E

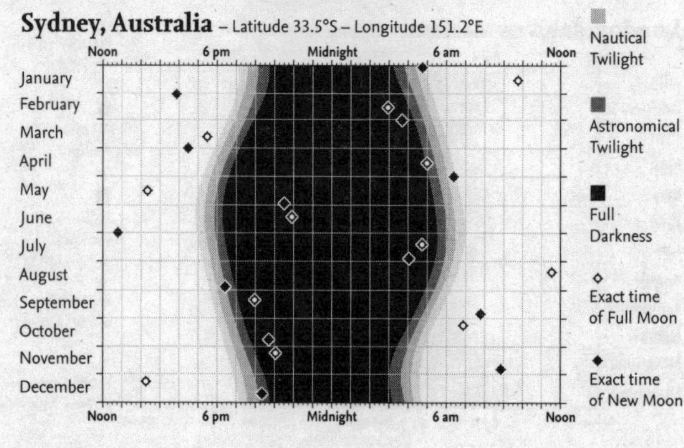

Tokyo, Japan – Latitude 35.7°N – Longitude 139.8°E

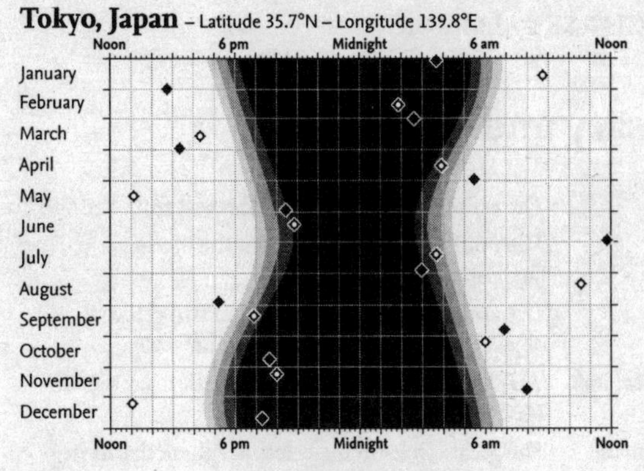

Washington, DC, USA – Latitude 38.9°N – Longitude 77.0°W

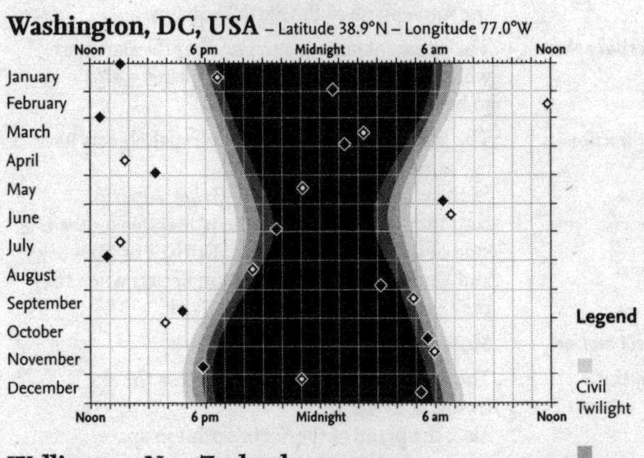

Wellington, New Zealand – Latitude 41.3°S – Longitude 174.8°E

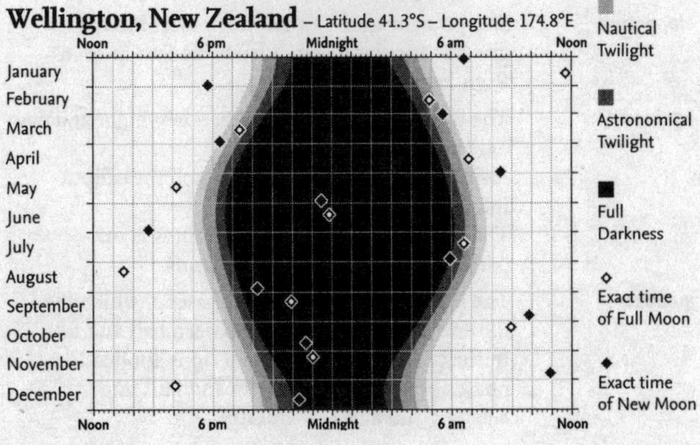

Glossary and Tables

aphelion	The point on an orbit that is farthest from the Sun.
apogee	The point on its orbit at which the Moon is farthest from the Earth.
appulse	The apparently close approach of two celestial objects; two planets, or a planet and star.
astronomical unit	(AU) The mean distance of the Earth from the Sun, 149,597,870 km.
celestial equator	The great circle on the celestial sphere that is in the same plane as the Earth's equator.
celestial sphere	The apparent sphere surrounding the Earth on which all celestial bodies (stars, planets, etc.) seem to be located.
conjunction	The point in time when two celestial objects have the same celestial longitude. In the case of the Sun and a planet, superior conjunction occurs when the planet lies on the far side of the Sun (as seen from Earth). For Mercury and Venus, inferior conjunction occurs when they pass between the Sun and the Earth.
direct motion	Motion from west to east on the sky.
ecliptic	The apparent path of the Sun across the sky throughout the year. Also: the plane of the Earth's orbit in space.
elongation	The point at which an inferior planet has the greatest angular distance from the Sun, as seen from Earth.
equinox	The two points during the year when night and day have equal duration. Also: the points on the sky at which the ecliptic intersects the celestial equator. The vernal (northern spring) equinox is of particular importance in astronomy.
gibbous	The stage in the sequence of phases at which the illumination of a body lies between half and full. In the case of the Moon, the term is applied to phases between First Quarter and Full, and between Full and Last Quarter.

inferior planet Either of the planets Mercury or Venus, which have orbits inside that of the Earth.

magnitude The brightness of a star, planet or other celestial body. It is a logarithmic scale, where larger numbers indicate fainter brightness. A difference of 5 in magnitude indicates a difference of 100 in actual brightness, thus a first-magnitude star is 100 times as bright as one of sixth magnitude.

meridian The great circle passing through the North and South Poles of a body and the observer's position; or the corresponding great circle on the celestial sphere that passes through the North and South Celestial Poles and also through the observer's zenith.

nadir The point on the celestial sphere directly beneath the observer's feet, opposite the zenith.

occultation The disappearance of one celestial body behind another, such as when stars or planets are hidden behind the Moon.

opposition The point on a superior planet's orbit at which it is directly opposite the Sun in the sky.

perigee The point on its orbit at which the Moon is closest to the Earth.

perihelion The point on an orbit that is closest to the Sun.

retrograde motion Motion from east to west on the sky.

superior planet A planet that has an orbit outside that of the Earth.

vernal equinox The point at which the Sun, in its apparent motion along the ecliptic, crosses the celestial equator from south to north. Also known as the First Point of Aries.

zenith The point directly above the observer's head.

zodiac A band, stretching 8° on either side of the ecliptic, within which the Moon and planets appear to move. It consists of 12 equal areas, originally named after the constellation that once lay within it.

The Constellations

There are 88 constellations covering the whole of the celestial sphere. The names themselves are expressed in Latin, and the names of stars are frequently given by Greek letters (see page 258) followed by the genitive of the constellation name or its three-letter abbreviation. The genitives, the official abbreviations and the English names of the various constellations are included.

Name	*Genitive*	*Abbr.*	*English name*
Andromeda	Andromedae	And	Andromeda
Antlia	Antliae	Ant	Air Pump
Apus	Apodis	Aps	Bird of Paradise
Aquarius	Aquarii	Aqr	Water Bearer
Aquila	Aquilae	Aql	Eagle
Ara	Arae	Ara	Altar
Aries	Arietis	Ari	Ram
Auriga	Aurigae	Aur	Charioteer
Boötes	Boötis	Boo	Herdsman
Caelum	Caeli	Cae	Burin
Camelopardalis	Camelopardalis	Cam	Giraffe
Cancer	Cancri	Cnc	Crab
Canes Venatici	Canum Venaticorum	CVn	Hunting Dogs
Canis Major	Canis Majoris	CMa	Big Dog
Canis Minor	Canis Minoris	CMi	Little Dog
Capricornus	Capricorni	Cap	Sea Goat
Carina	Carinae	Car	Keel
Cassiopeia	Cassiopeiae	Cas	Cassiopeia
Centaurus	Centauri	Cen	Centaur
Cepheus	Cephei	Cep	Cepheus
Cetus	Ceti	Cet	Whale
Chamaeleon	Chamaeleontis	Cha	Chameleon
Circinus	Circini	Cir	Compasses
Columba	Columbae	Col	Dove
Coma Berenices	Comae Berenices	Com	Berenice's Hair
Corona Australis	Coronae Australis	CrA	Southern Crown
Corona Borealis	Coronae Borealis	CrB	Northern Crown

Name	*Genitive*	*Abbr.*	*English name*
Corvus	Corvi	Crv	Crow
Crater	Crateris	Crt	Cup
Crux	Crucis	Cru	Southern Cross
Cygnus	Cygni	Cyg	Swan
Delphinus	Delphini	Del	Dolphin
Dorado	Doradus	Dor	Dorado
Draco	Draconis	Dra	Dragon
Equuleus	Equulei	Equ	Little Horse
Eridanus	Eridani	Eri	River Eridanus
Fornax	Fornacis	For	Furnace
Gemini	Geminorum	Gem	Twins
Grus	Gruis	Gru	Crane
Hercules	Herculis	Her	Hercules
Horologium	Horologii	Hor	Clock
Hydra	Hydrae	Hya	Water Snake
Hydrus	Hydri	Hyi	Lesser Water Snake
Indus	Indi	Ind	Indian
Lacerta	Lacertae	Lac	Lizard
Leo	Leonis	Leo	Lion
Leo Minor	Leonis Minoris	LMi	Little Lion
Lepus	Leporis	Lep	Hare
Libra	Librae	Lib	Scales
Lupus	Lupi	Lup	Wolf
Lynx	Lyncis	Lyn	Lynx
Lyra	Lyrae	Lyr	Lyre
Mensa	Mensae	Men	Table Mountain
Microscopium	Microscopii	Mic	Microscope
Monoceros	Monocerotis	Mon	Unicorn
Musca	Muscae	Mus	Fly
Norma	Normae	Nor	Set Square
Octans	Octantis	Oct	Octant
Ophiuchus	Ophiuchi	Oph	Serpent Bearer
Orion	Orionis	Ori	Orion
Pavo	Pavonis	Pav	Peacock
Pegasus	Pegasi	Peg	Pegasus
Perseus	Persei	Per	Perseus

Name	*Genitive*	*Abbr.*	*English name*
Phoenix	Phoenicis	Phe	Phoenix
Pictor	Pictoris	Pic	Painter's Easel
Pisces	Piscium	Psc	Fishes
Piscis Austrinus	Piscis Austrini	PsA	Southern Fish
Puppis	Puppis	Pup	Stern
Pyxis	Pyxidis	Pyx	Compass
Reticulum	Reticuli	Ret	Net
Sagitta	Sagittae	Sge	Arrow
Sagittarius	Sagittarii	Sgr	Archer
Scorpius	Scorpii	Sco	Scorpion
Sculptor	Sulptoris	Scu	Sculptor
Scutum	Scuti	Sct	Shield
Serpens	Serpentis	Ser	Serpent
Sextans	Sextantis	Sex	Sextant
Taurus	Tauri	Tau	Bull
Telescopium	Telescopii	Tel	Telescope
Triangulum	Trianguli	Tri	Triangle
Triangulum Australe	Trianguli Australis	TrA	Southern Triangle
Tucana	Tucanae	Tuc	Toucan
Ursa Major	Ursae Majoris	UMa	Great Bear
Ursa Minor	Ursae Minoris	UMi	Lesser Bear
Vela	Velorum	Vel	Sails
Virgo	Virginis	Vir	Virgin
Volans	Volantis	Vol	Flying Fish
Vulpecula	Vulpeculae	Vul	Fox

The Greek Alphabet

α	Alpha	ι	Iota	ρ	Rho
β	Beta	κ	Kappa	σ (ς)	Sigma
γ	Gamma	λ	Lambda	τ	Tau
δ	Delta	μ	Mu	υ	Upsilon
ε	Epsilon	ν	Nu	ϕ (φ)	Phi
ζ	Zeta	ξ	Xi	χ	Chi
η	Eta	ο	Omicron	ψ	Psi
θ (ϑ)	Theta	π	Pi	ω	Omega

Asterisms
Apart from the constellations (88 of which cover the whole sky), listed on pages 256–258, certain groups of stars, which may form a small part of a larger constellation, are readily recognizable and have been given individual names. These groups are known as ***asterisms***, and the most famous (and well-known) is the 'Plough' or 'Big Dipper', the common name for the seven brightest stars in the constellation of Ursa Major, the Great Bear. The names and details of some asterisms mentioned in this book are given in this list.

Some common asterisms	
Belt of Orion	δ, ε and ζ Orionis
Big Dipper	α, β, γ, δ, ε, ζ and η Ursae Majoris
Cat's Eyes	λ and υ Scorpii
Circlet	γ, θ, ι, λ and κ Piscium
False Cross	ε and ι Carinae and δ and κ Velorum
Fish Hook	α, β, δ and π Scorpii
Guards (or Guardians)	β and γ Ursae Minoris
Head of Cetus	α, γ, ξ^2, μ and λ Ceti
Head of Draco	β, γ, ξ and ν Draconis
Head of Hydra	δ, ε, ζ, η, ρ and σ Hydrae
Job's Coffin	α, β, γ and δ Delphini
Keystone	ε, ζ, η and π Herculis
Kids	ε, ζ and η Aurigae
Little Dipper	β, γ, η, ζ, ε, δ and α Ursae Minoris
Lozenge	= Head of Draco
Milk Dipper	ζ, γ, σ, φ and λ Sagittarii
Plough	= Big Dipper
Pointers	α and β Ursae Majoris
Pot	= Saucepan
Saucepan	ι, θ, ζ, ε, δ and η Orionis
Sickle	α, η, γ, ζ, μ and ε Leonis
Southern Pointers	α and β Centauri
Square of Pegasus	α, β and γ Pegasi with α Andromedae
Sword of Orion	θ and ι Orionis
Teapot	γ, ε, δ, λ, φ, σ, τ and ζ Sagittarii
Wain (or Charles' Wain)	= Big Dipper
Water Jar	γ, η, κ and ζ Aquarii
Y of Aquarius	= Water Jar

107 Named stars brighter than magnitude 2.75

Name	*Con*	*Mag*	*Name*	*Con*	*Mag*
Achernar	α Eri	0.45	**Aspidiske**	ι Car	2.21
Acrab	β Sco	2.56	**Athebyne**	η Dra	2.73
Acrux	α Cru	0.77	**Atria**	α TrA	1.91
Adhara	ε CMa	1.50	**Avior**	ε Car	1.86
Aldebaran	α Tau	0.87	**Bellatrix**	γ Ori	1.64
Alderamin	α Cep	2.45	**Betelgeuse**	α Ori	0.45
Algieba	γ Leo	2.01	**Canopus**	α Car	-0.62
Algol	β Per	2.09	**Capella**	α Aur	0.08
Alhena	γ Gem	1.93	**Caph**	β Cas	2.28
Alioth	ε UMa	1.76	**Castor**	α Gem	1.58
Aljanah	ε Cyg	2.48	**Deneb**	α Cyg	1.25
Alkaid	η UMa	1.85	**Denebola**	β Leo	2.14
Almach	γ And	2.10	**Diphda**	β Cet	2.04
Alnair	α Gru	1.73	**Dschubba**	δ Sco	2.29
Alnilam	ε Ori	1.69	**Dubhe**	α UMa	1.81
Alnitak	ζ Ori	1.74	**Elnath**	β Tau	1.65
Alphard	α Hya	1.99	**Eltanin**	γ Dra	2.24
Alphecca	α CrB	2.22	**Enif**	ε Peg	2.38
Alpheratz	α And	2.07	**Fomalhaut**	α PsA	1.17
Alsephina	δ Vel	1.93	**Gacrux**	γ Cru	1.59
Altair	α Aql	0.76	**Gienah**	γ Crv	2.58
Aludra	η CMa	2.45	**Hadar**	β Cen	0.61
Ankaa	α Phe	2.40	**Hamal**	α Ari	2.01
Antares	α Sco	1.06	**Hassaleh**	ι Aur	2.69
Arcturus	α Boo	-0.05	**Izar**	ε Boo	2.35
Arneb	α Lep	2.58	**Kaus Australis**	ε Sgr	1.79
Ascella	ζ Sgr	2.60	**Kaus Media**	δ Sgr	2.72

Name	*Con*	*Mag*	*Name*	*Con*	*Mag*
Kochab	β UMi	2.07	**Procyon**	α CMi	0.40
Kraz	β Crv	2.65	**Rasalhague**	α Oph	2.08
Larawag	ε Sco	2.29	**Regulus**	α Leo	1.36
Lesath	υ Sco	2.70	**Rigel**	β Ori	0.18
Mahasim	ϑ Aur	2.65	**Rigil Kentaurus**	α Cen	-0.29
Markab	α Peg	2.49	**Ruchbah**	δ Cas	2.66
Markeb	κ Vel	2.47	**Sabik**	η Oph	2.43
Menkalinan	β Aur	1.90	**Sadr**	γ Cyg	2.23
Menkar	α Cet	2.54	**Saiph**	κ Ori	2.07
Menkent	ϑ Cen	2.06	**Sargas**	ϑ Sco	1.86
Merak	β UMa	2.34	**Scheat**	β Peg	2.44
Miaplacidus	β Car	1.67	**Schedar**	α Cas	2.24
Mimosa	β Cru	1.25	**Shaula**	λ Sco	1.62
Mintaka	δ Ori	2.25	**Sheratan**	β Ari	2.64
Mirach	β And	2.07	**Sirius**	α CMa	-1.44
Mirfak	α Per	1.79	**Spica**	α Vir	0.98
Mirzam	β CMa	1.98	**Suhail**	λ Vel	2.23
Mizar	ζ UMa	2.23	**Tarazed**	γ Aql	2.72
Muphrid	η Boo	2.68	**Tiaki**	β Gru	2.07
Naos	ζ Pup	2.21	**Unukalhai**	α Ser	2.63
Nunki	σ Sgr	2.05	**Vega**	α Lyr	0.03
Peacock	α Pav	1.94	**Wezen**	δ CMa	1.83
Phact	α Col	2.65	**Yed Prior**	δ Oph	2.73
Phecda	γ UMa	2.41	**Zosma**	δ Leo	2.56
Polaris	α UMi	1.97	**Zubenelgenubi**	α Lib	2.75
Pollux	β Gem	1.16	**Zubeneschamali**	β Lib	2.61
Porrima	γ Vir	2.74			

Further Information

Books

Bone, Neil (1993), ***Observer's Handbook: Meteors***, George Philip, London & Sky Publishing Corp., Cambridge, Mass.

Cook, J., ed. (1999), ***The Hatfield Photographic Lunar Atlas***, Springer-Verlag, New York

Dunlop, Storm (2006), ***Wild Guide to the Night Sky***, Harper Perennial, New York & Smithsonian Press, Washington DC

Dunlop, Storm (2012), ***Practical Astronomy***, 2nd edn, Firefly, Buffalo

Dunlop, Storm, Rükl, Antonin & Tirion, Wil (2005), ***Collins Atlas of the Night Sky***, HarperCollins, London & Smithsonian Press, Washington DC

Grego, Peter (2016), ***Moon Observer's Guide***, Firefly, Richmond Hill

Heifetz, Milton D. & Tirion, Wil (2017), ***A Walk through the Heavens***, 4th edn, Cambridge University Press, Cambridge

Mellinger, Axel & Hoffmann, Susanne (2005), ***The New Atlas of the Stars***, Firefly, Richmond Hill

O'Meara, Stephen J. (2008), ***Observing the Night Sky with Binoculars***, Cambridge University Press, Cambridge

Pasachoff, Jay M. (1999), ***Peterson Field Guides: Stars and Planets***, 4th edn, Houghton Mifflin, Boston

Ridpath, Ian, ed. (2003), ***Oxford Dictionary of Astronomy***, 2nd edn, Oxford University Press, Oxford & New York

Ridpath, Ian, ed. (2004), ***Norton's Star Atlas***, 20th edn, Pi Press, New York

Ridpath, Ian (2018), ***Star Tales***, 2nd edn, Lutterworth Press, Cambridge UK

Ridpath, Ian & Tirion, Wil (2004), ***Collins Gem – Stars***, HarperCollins, London

Ridpath, Ian & Tirion, Wil (2017), ***Collins Pocket Guide Stars and Planets***, 5th edn, HarperCollins, London

Ridpath, Ian & Tirion, Wil (2012), ***Monthly Sky Guide***, 10th edn, Dover Publications, New York

Rükl, Antonín (1990), ***Hamlyn Atlas of the Moon***, Hamlyn, London & Astro Media Inc., Milwaukee

Rükl, Antonín (2004), ***Atlas of the Moon***, Sky Publishing Corp., Cambridge, Mass.

Scagell, Robin (2014), ***Stargazing with a Telescope***, Firefly, Richmond Hill

Scagell, Robin (2015), ***Firefly Complete Guide to Stargazing***, Firefly, Richmond Hill

Scagell, Robin & Frydman, David (2014), ***Stargazing with Binoculars***, Firefly, Richmond Hill

Sky & Telescope (2017), ***Astronomy 2018***, Sky Publishing Corp., Cambridge, Mass.

Stimac, Valerie (2019), ***Dark Skies: A Practical Guide to Astrotourism***, Lonely Planet

Tirion, Wil (2011), ***Cambridge Star Atlas***, 4th edn, Cambridge University Press, Cambridge

Tirion, Wil & Sinnott, Roger (1999), ***Sky Atlas 2000.0***, 2nd edn, Sky Publishing Corp., Cambridge, Mass. & Cambridge University Press, Cambridge

Journals

Astronomy, Astro Media Corp., 21027 Crossroads Circle, P.O. Box 1612, Waukesha, WI 53187-1612.
http://www.astronomy.com

Sky & Telescope, Sky Publishing Corp., Cambridge, MA 02138-1200.
http://www.skyandtelescope.com/

Societies

American Association of Variable Star Observers (AAVSO), 49 Bay State Rd., Cambridge, MA 02138. Although primarily concerned with variable stars, the AAVSO also has a solar section.

American Astronomical Society (AAS), 1667 K Street NW, Suite 800, Washington, DC 20006, New York.
http://aas.org/

American Meteor Society (AMS), Geneseo, New York.
http://www.amsmeteors.org/

Association of Lunar and Planetary Observers (ALPO), ALPO Membership Secretary/Treasurer, P.O. Box 13456, Springfield, IL 62791-3456.
An organization concerned with all forms of amateur astronomical observation, not just the Moon and planets, with numerous coordinated observing sections.
http://alpo-astronomy.org/

Astronomical League (AL), 9201 Ward Parkway Suite #100,
Kansas City, MO 64114.
An umbrella organization consisting of over 240 local amateur astronomical societies across the United States.
https://www.astroleague.org/

British Astronomical Association (BAA), Burlington House, Piccadilly,
London W1J 0DU.
The principal British organization (but with a worldwide membership) for amateur astronomers (with some professional members), particularly for those interested in carrying out observational programs.
http://www.britastro.org/

International Meteor Organization (IMO)
An organization coordinating observations of meteors worldwide.
http://www.imo.net/

Royal Astronomical Society of Canada (RASC), 203 – 4920 Dundas Street W.,
Toronto, ON M9A 1B7.
The principal Canadian astronomical organization, with both professional and amateur members. It has 28 local centres.
http://rasc.ca/

Software

Planetary, Stellar and Lunar Visibility (planetary and eclipse freeware):
Alcyone Software, Germany.
http://www.alcyone.de

Redshift, Redshift-Live.
http://www.redshift-live.com/en/

Starry Night & Starry Night Pro, Sienna Software Inc., Toronto, Canada.
http://www.starrynight.com

Internet sources

There are numerous sites about all aspects of astronomy, and all have numerous links. Although many amateur sites are excellent, treat any statements and data with caution. The sites listed below offer accurate information. Please note that the URLs may change. If so, use a good search engine, such as Google, to locate the information source.

Information

Astronomical data (inc. eclipses) HM Nautical Almanac Office:
http://astro.ukho.gov.uk

Auroral information Michigan Tech:
http://www.geo.mtu.edu/weather/aurora/

Comets JPL Solar System Dynamics:
http://ssd.jpl.nasa.gov/

Deep-sky objects Saguaro Astronomy Club Database:
http://www.virtualcolony.com/sac/

Eclipses NASA Eclipse Page:
http://eclipse.gsfc.nasa.gov/eclipse.html

Moon (inc. Atlas) Inconstant Moon:
http://www.inconstantmoon.com/

Planets Planetary Fact Sheets:
http://nssdc.gsfc.nasa.gov/planetary/planetfact.html

Satellites (inc. International Space Station)
Heavens Above: http://www.heavens-above.com/
Visual Satellite Observer: http://www.satobs.org/

Star Chart
http://www.skyandtelescope.com/observing/interactive-sky-watching-tools/interactive-sky-chart/

What's Visible
Skyhound: http://www.skyhound.com/sh/skyhound.html
Skyview Cafe: http://www.skyviewcafe.com

Institutes and Organizations

European Space Agency: http://www.esa.int/

International Dark-Sky Association: http://www.darksky.org/

RASC Dark Sky: https://www.rasc.ca/dark-sky-site-designations

Jet Propulsion Laboratory: http://www.jpl.nasa.gov/

Lunar and Planetary Institute: http://www.lpi.usra.edu/

National Aeronautics and Space Administration: http://www.hq.nasa.gov/

Solar Data Analysis Center: http://umbra.gsfc.nasa.gov/

Space Telescope Science Institute: http://www.stsci.edu/

Acknowledgements

23	NASA/GSFC/SVS
48	Stephen Pitt
56	Denis Buczymski
57	NASA
67	Natural History Museum, London
75	National Museum of Natural History, Washington
85	ESA/Rosetta/NAVCAM
99	OKB-1
100	NASA/Lunar Reconnaissance Orbiter
108	Peter Komka/EPA-EFE/Shutterstock
131	Duncan Waldron, Brisbane, Australia
138	National Maritime Museum, Greenwich, London
147	NASA/Wikipedia
149	NASA / Jet Propulsion Laboratory / University of Maryland
163	NASA/JPL
164	NASA/JPL/JHUAPL
165	NASA
172	NASA/JPL
173	NASA/JPL-Caltech/UCAL/MPS/DLR/IDA
183	NASA
183	NASA (from GPN-2000-001623)
197	Duncan Waldron, Brisbane, Australia
227	NASA/Lunar Orbiter 5
228	NASA/LRO

The authors would also like to thank Barry Hetherington.

Specialist editorial support was provided by Hannah Banyard, Public Astronomy Officer and Dr Gregory Brown, Public Astronomy Officer at the Royal Observatory, part of Royal Museums Greenwich.

Index

EXPLORE OUR RANGE OF ASTRONOMY TITLES

Other titles by Storm Dunlop and Wil Tirion

2022 Guide to the Night Sky: Britain and Ireland
978-0-00-839353-3

2022 Guide to the Night Sky: North America
978-0-00-846986-3

2022 Guide to the Night Sky: Southern Hemisphere
978-0-00-846980-1
Latest editions of our bestselling month-by-month guides for exploring the skies. These guides are an easy introduction to astronomy and a useful reference for seasoned stargazers.

Collins Planisphere | 978-0-00-754075-4
Easy-to-use practical tool to help astronomers to identify the constellations and stars every day of the year. For latitude 50°N, suitable for use anywhere in Britain and Ireland, Northern Europe, Canada and Northern USA

Also available

Astronomy Photographer of the Year: Collection 10
978-0-00-846987-0
Winning and shortlisted images from the 2021 Insight Investment Astronomy Photographer of the Year competition, hosted by the Royal Observatory, Greenwich. The images include aurorae, galaxies, our Moon, our Sun, people and space, planets, comets and asteroids, skyscapes, stars and nebulae.

Stargazing | 978-0-00-819627-1
The prefect manual for beginners to astronomy – introducing the world of telescopes, planets, stars, dark skies and celestial maps.

Moongazing | 978-0-00-830500-0
An in-depth guide for all aspiring astronomers and Moon observers, with detailed Moon maps. Covers the history of lunar exploration and the properties of the Moon, its origin and orbit.

The Moon | 978-0-00-828246-2
A celebration of our celestial neighbour, exploring people's fascination with our only natural satellite, illuminating how art and science meet in our profound connection with the Moon.

Northern Lights | 978-0-00-846555-1
Discover the incomparable beauty of the Northern Lights with this accessible guide for both aspiring astronomers and seasoned night sky observers alike.